The Emerging New Humanity

NOW Is The Time

The Emerging New Humanity

NOW Is The Time

Amid political gridlock, global crises and growing terrorism are surprising trends of a brighter future

Robert B. Calkins

Published by Robert B. Calkins
Issaquah, WA

Printed by CreateSpace
Copyright © 2016 by Robert B. Calkins
All right reserved, including the right to reproduce this book or portions thereof in any form whatsoever.

Edition 1

Calkins, Robert B.
Science, Philosophy, Future, Evolution, Change, Chaos, Humanity, Transformation, Cultural Development, Environment, Global Climate Change, Globalization, Hunger, Civilization, Revolutionary, Nature.

ISBN-13: 978-1539019916

Cover art and design by Robert B. Calkins

CONTENTS

Preface and Acknowledgments................................. vii

Part I TODAY'S CHANGING WORLD   1

    1. What's Happening?................................. 3
    2. Why the Problems?............................... 21
    3. Origins of our Current Culture..................... 33

Part II A SCIENTIFIC VIEW   45

    4. Recent Developments in Science................... 47
    5. How Change Happens............................. 63
    6. The Need for Today's Chaos....................... 71

Part III AN INTEGRAL VIEW   79

    7. Integrated Systems Thinking....................... 81
    8. Evolution, the Big Picture........................... 93
    9. Where We Are Now................................ 111

Part IV A CHANGE OF PACE   125

    10. The Accelerating Pace of Change................. 127
    11. Where We Are Going............................. 139

Part V WHERE DO WE GO FROM HERE?   153

    12. The Promise of Now.............................. 155
    13. Riding the Wave of Change 169

Epilogue.. 177
Glossary... 181
Notes.. 191
Index.. 205
About the Author... 213

Preface

READ THIS FIRST, at least this paragraph. Some people read the end of a book first or skip around. This particular book is best read from start to finish in order, because some of the later material may not make sense until the foundation is laid. I wrote this book by building up a group of ideas that lead to an integrated picture. Reading it out of order may be confusing. Of course the choice is yours.

We all have wondered about events and conditions in our modern world. Things are moving so fast, there is much turmoil, chaos and change. Is there no hope for peace? It seems the pace of change is faster and faster, as if it can't continue speeding up like this. Where will it all end? There does not seem to be any real purpose to it all. Many of us are getting frustrated at the seeming lack of progress or leadership in important matters and are looking for answers, hoping to find some peace of mind.

This book is about why things are the way they are and where they are going. In some sense, things are just where they need to be. There is hope for peace.

This journey starts with a look at the problems we face today and some of the underlying thinking that helps feed them, followed by some of the pertinent newer scientific ideas and fields and how they relate to life, civilization and our future. My method is a step-by-step approach to add meaning to the seeming chaos, and provide a broad road map of where it is leading and why some of the unpleasant things are really okay. I use a systems approach to look at the big picture, without trying to solve individual problems. My rationale uses science, not wishful thinking, to examine things.

I hope after reading this book, you will have a better feeling about some of the happenings you see occurring in the world and your life. You may also be in a position to make more informed decisions.

There are five parts to this book. In Part I, we look at today's changing world. In Part II, we examine this world from a scientific point of view. Part III examines things from an integrated theory perspective. Part IV discusses some of the results of these changes, and finally, Part V takes a look at where we are heading.

In some cases, dates are used. Many of these dates are approximate and still in debate in the scientific community. In order to not bore you with endless "approximately" and ranges of dates, one date has been selected that I feel best represents the topic. In other words, the exact date is subject to debate and to change without notice. The book was up-to-date when I wrote it, but things were changing even as I was writing. I have already had to change several things when new information was published. Exact dates do not change the basic ideas presented. Words in **bold** are defined in the glossary.

Several ideas in this book are not universally accepted by all of science, but then almost nothing in science is. Science is a constantly changing field of knowledge and ideas. What is true today is tomorrow's folly. It has happened in science that truth became folly and later "truth" again. It has also happened many times that one day's folly later became truth. I leave it up to the reader to decide which this book is. If you decide it is folly, read it again in a year or two. Things are changing that fast.

I would like to thank and acknowledge the life support and editing of my wonderful wife and partner, Dana Ericson (with no "k"), without which this book would be semi-readable. I also would like to acknowledge the editing support of Anne Bock, Arnie Marcus, Jan Viney, Dan Lyons and David A. Pardo, PhD.

Part I

Today's Changing World

1

Don't tell your problems to people: eighty percent don't care; and the other twenty percent are glad you have them.
—Lou Holtz

What's Happening?

More than a decade into the twenty-first century, mankind stands at a crossroads. The direction we have been taking for the last four hundred years, since the **Renaissance**, is leading to a dead end. The result is a world in turmoil. Multiple global crises are already in place and getting worse with no coherent vision or leadership on how to solve them. Albert Einstein said "The significant problems we face cannot be solved at the same level of thinking we were at when we created them"[1]. That seems to be the case today. But another path has been opening up. Slowly, in the last third of the twentieth century it started to form. Today around the world we are witnessing a grassroots cultural change that holds the promise of solving these problems at a different level of thinking than they were created at.

As we look at the world today, we see amongst the rapid change, growing chaos and an increasing sense of something wrong with the way things are. As we shall see, this is actually a good thing. The old way of doing things has led humanity on a course of global disaster. In some ways, it was a necessary course, but its time has passed. We are now in a major transition to a new order and understanding. That kind of change does not happen smoothly.

History shows that the kind of real change needed does not come without turmoil and struggle, and science can now explain why this is so. It is not an easy time, but if understood, it can be looked upon with anticipation, hope and wonder, even when it looks like all is lost. Science is undergoing this transition just as the rest of society is. We will see how this change is being led by new discoveries and new scientific disciplines, and how they show the light at the end of our current tunnel; a very bright light.

Hold on to your hat, it's a wild ride, but one that must be taken. If you know where it's going, the ride will be a lot smoother than if all you know is that it's wild and looks like it will get worse.

There is a myth of an old Chinese saying, meant as a curse, "May you live in interesting times". We are, but it ends up being a blessing. Science can't tell us the future, but it can explain the process and where it is generally headed. That is well worth knowing when it seems there is only disorder and decay in the immediate present. This book will not give the details of the future, but it is meant as an overview of the big picture of where we are going; more a global map rather than a street map.

In part I, we briefly review the current state of affairs and some reasons why it is the way it is. You may feel you already know about today's world. However, it is important to look at where we are and how we got here for two reasons. First, we spend most of our attention on what is happening now and what is happening in our own life. This is not only natural but to some degree unavoidable. We tend to develop our own local, current sense of reality. Taking a step back and looking globally at the bigger picture and wider time frame can broaden our perspective and allow us to understand things that may otherwise seem odd or nonsensical.

Secondly, revisiting current affairs helps establish a common base of understanding so we are on the same page, metaphorically speaking.

In today's complex world, it seems the significant problems have been getting bigger, not smaller. First the bad news: in the rest of this chapter we take a quick look at some of the bigger problems. The point here is not to sound as if doomsday is at hand, although it may appear that way for a time. Please don't stop here, as the rest of this book has a much more positive outlook.

Global Climate Change

Global Climate Change is considered to be a clear fact by the vast majority of scientists and scientific organizations today[2]. It is possible some of the change may be due to natural causes, but there is widespread scientific consensus on most of it being caused by human made **greenhouse gases**. See figures 1.1 and 1.2. What is noticeable now is that a number of factors seem to be occurring faster than any of the models predicted. For decades warnings about the adverse effects of global climate change have been made. Today, those effects are starting to show up.

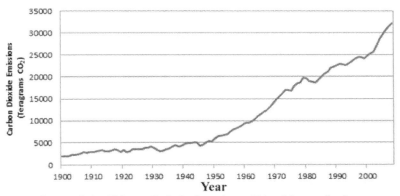

Figure 1.1. Rise of global carbon Dioxide emissions over past 200 years[22].

One reason it has taken so long to be generally recognized is that some aspects of big corporate interests with vast sources of money have attempted to obscure it[3, 4]. One tactic has been to put up a smoke screen by funding agreeable scientists to do studies that downplay climate change. Another has been the successful use of advertising spending power to stop news stories or intimidate what is said on television programming. Another tactic has been to use think tanks that get funding from anti-climate change interests to put out contrary studies that cloud the truth. A fourth tactic has been to use political power to stifle legitimate scientific opinions within the government. It is little wonder that many people were slow to realize the full danger of global climate change.

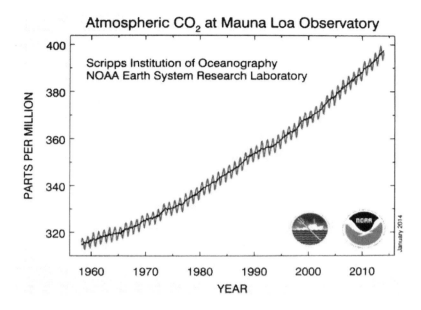

Figure 1.2. Atmospheric carbon dioxide rise during past 50 years[23].

Land is already being lost to the rise in ocean levels. Melting permafrost is damaging buildings and releasing more greenhouse gases. Climate is changing growing seasons, causing pests to invade new territories, and expanding desert areas. More severe weather has been predicted and is being experienced with stronger hurricanes, tornadoes, droughts, floods and wildfires. As of 2009 there were 25 million environmental refugees. This number is expected to double in short order and in 40 years, to top 200 million people[5].

Global climate change will increase the world's arid lands that can't support adequate farming. It will flood prime agricultural low lands in many parts of the world. It will also disrupt society through increasingly severe floods and droughts, while making fresh water scarcer[5].

As the warmer climates move further northward from the tropics, tropical diseases such as malaria and Lyme disease are following. A host of environmental pests on land and in the sea will also move north. One such pest may be the pine beetles that are ravishing more than 37-million acres of forest from Yosemite National Park in California all the way north into southern Canada. Several days of sub twenty degree weather normally kills pine beetles, but a string of years of unusually warm winters has led to an explosion of the tree killing beetles. Wide spread tree kill-offs then lead to record fire storms in summer, floods, and more carbon dioxide[6].

Global climate change models are showing that some of the changes take a century or more to start having an effect. They also show that if we stopped adding greenhouse gasses to the atmosphere right now, it would take a century or two for the global climate to fully react, and global climate change would continue to worsen for some time[2].

More recent data have surprised scientists by showing that of the 111 Greenland glaciers surveyed using NASA satellites and lasers, 81 are melting at an accelerating rate, termed "runaway melt mode". See figure 1.3.

One scientist has called the findings "alarming"[7]. This discovery is significant because, prior to it, the melting ice of Greenland and Antarctica was not considered in global climate models, mainly because it was not well understood. Once these studies were completed, the implication for the rise of sea level was significant[8]. As a result the predicted sea level rise during the 21st century increased by a factor of six. This increase means the coastal flooding potential is much higher that previously stated.

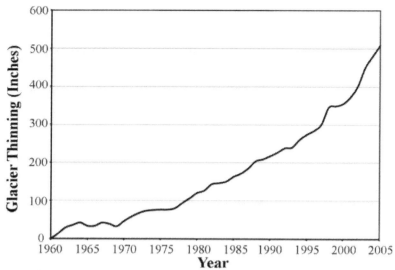

Figure 1.3. Average Cumulative Thinning of Worlds Glaciers in the past 45 years.

In the San Francisco, California area for instance, this could mean the loss of San Francisco and Oakland international airports, one-hundred-forty schools, twenty-nine wastewater treatment facilities, two-hundred-eighty miles of rail roads and over three-thousand- five hundred miles of paved roads[9]. The southern portion of Florida, including the Miami area would also be lost, as would large areas of coastal lands and communities elsewhere around the country. Flooding in Miami Beach during high tides is now already a problem.

With adverse effects of global climate change now happening, one might think the political leaders of the world would be hard at work solving this problem. However, much of our political life is ruled by influence and money. Big business is not interested in spending considerable money to solve this problem. However, we should not be too quick in blaming our leaders and big business. A large part of our population is ignoring the problem. It is such a big problem and we can't see what we can do, so we put it out of mind. If it is not a hot topic, the politicians tend to put it out of mind also. The results of global climate change are a human created disaster. That disaster is now fast approaching and we have had ample warning. If we act too late, the toll may be monumental. It may take centuries to change once we pass the point where things go really wrong.

Environmental Problems

Environmental problems have been mounting at an accelerating rate. Some are tied in with global warming. Others include air, land and water pollution, as well as topsoil erosion, diminishing of forests, scarcity of water, and spread of pests and diseases.

Pollution has been a growing problem. Air pollution in cities has been linked to higher death rates. Exposure over a long time to common levels of urban air pollution in America has been associated with a twelve percent increase of cardiovascular mortality[10]. This is just one risk factor for air pollution. In parts of Asia, entire regions, not just urban areas, now have unhealthy air quality for much of the summer months. The entire Eastern half of the United States has a persistent summer haze of air pollution. The US Environmental Protection Agency stated that "tens of thousands of people die each year from breathing tiny particles in the environment"[6].

The list of banned or dangerous environmental substances continues to grow. Past banned substances have a long list, including DDT, PCB, dioxins and light chromium-6. More and more chemical substances enter our use and environment. Only years later are some discovered to be very dangerous and banned or limited. Most of these chemicals do not occur in nature. Heavy metals, herbicides, pesticides, chlorinated hydrocarbons, hormones and medications (having passed through a human or animal by way of their waste and into the environment) are all problems for the environment and for us. Four-hundred-million tons of hazardous waste are generated each year world wide[11].

The seemingly endless oceans of the world have long been a dumping place for waste. But even the vast oceans are clogging up. There is a huge area of the mid-Pacific Ocean that contains forty-six thousand pieces of plastic floating on every square mile. Another two-and-a-half-million pieces enter the world's oceans every hour[12]. It is estimated that eighteen **billion** disposable diapers end up in the oceans each year. There are also tens of thousands of miles of monofilament drift nets and fishing lines floating in the oceans, as well as all kinds of junk including ropes, bottles, toys, motor-oil jugs, and tires. Pollution rates are growing with no signs of slowing down. All the plastic that enters the ocean stays in the ocean. It may break up, but it's still there as plastic. Water samples from various parts of our oceans have plastic particles in them.

Marine researchers have provided evidence of plastic contamination of wildlife on the molecular level[12]. Plastic particles have been found to outnumber plankton by a ratio of six to one. The problem here is that plankton is at the bottom of the food chain. If pollutants are entering the bottom of the food chain at the molecular level, they will be affecting the entire food chain all the way back up to the humans that made and discarded the plastic.

Acidification of our oceans is another pollution problem and is similar to global warming. As the levels of atmospheric carbon dioxide rise, they increase in the oceans washed from the atmosphere by rain. Carbon dioxide in water produces carbonic acid[13]. As with other factors, new research indicates acidification is occurring much faster than climate models have predicted. Acidification can have a major effect on shifting the balance of the oceans ecosystem, wreaking havoc on marine life.

Pollution of drinking water in the US is a growing problem[14]. In one year, over five hundred thousand violations of water pollution laws occurred, but the vast majority of polluters go free. Pollution of drinking waters includes chemicals, metals, drugs, herbicides, pesticides, and hormones. It's not so much whether drinking water is contaminated, but is it safe. Many chemicals and especially combinations of chemicals do not have safety levels determined.

The problem of massive environmental pollution continues to grow worldwide. The old "solution" was to dilute hazardous substances to an acceptable level. But with more sources and more types of pollution entering the picture all the time, that "solution" is not doing the job.

Mass Extinction of Species

Science recognizes five past major occurrences of mass extinction of life forms sense the beginning of life on Earth. These have been identified through fossil records. Causes of all of these are still in debate, but include significant atmospheric disruptions due to asteroid impacts and massive volcanism episodes. Many scientists today believe we are now seeing the sixth mass extinction. The difference with this one is that it is human caused with climate change the significant driver[15]. Figure 1.4 shows the trend in extinctions as compared to what would be expected through normal evolution. While this may not seem as big a problem as some issues, the biodiversity of life is part of our environment and it is not a healthy sign when we seem to be entering a major mass extinction episode caused by the same environment we live in.

Population Explosion

In the twentieth Century, the world population quadrupled. Experts say that, as the twenty-first century started, there were almost six-billion (6,000,000,000) people on earth. See figure 1.5. These people consumed 1.4 times as much as the earth could sustain in the long run[5]. It is not mathematically possible to quadruple again this century given the limitations in what nature and the earth can provide.

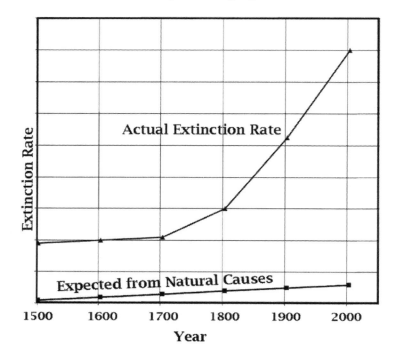

Figure 1.4 Current Mass Extinction

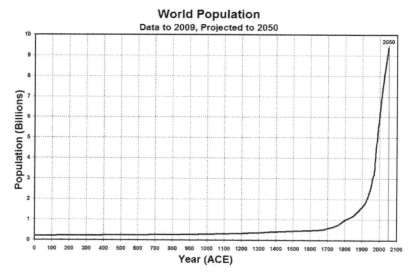

Figure 1.5. The Population Explosion

What's Happening?

The population explosion will be stopped, it's just a case of how – famine, pestilence, war, a change in culture, or a mixture of these. One thing is certain, things will change. The question is how and how painful will the changes be? Will the changes happen by skillful management by world governments, by an aware citizenry, or by struggle and disaster? At the rate world governments are currently dealing with this problem, the prospects do not look good.

The basic reason there has been a population explosion is that life expectancy has gone up and infant mortality has gone down while birth rates have continued relatively unchanged in much of the world. The rate of increase has been slowing in the developed world with lower birth rates and a leveling out of the growth in life expectancy. World average life expectancy at birth in the period from 1800 to 2000 went from 30 to 67[16]. The population of the US is forecast to double during the twenty-first century.

The biggest problem with population growth is that the largest growth occurs in the less developed countries, those which can least support the growth. The threat of starvation and social unrest due to population pressures is real.

Diminishing Resources

With the world population now having passed seven billion (7,000,000,000), the question is how can we sustain the living style we have become accustomed to? There is now abundant scientific evidence that we are already living well beyond long term **sustainability**[5,17]. A 2008 study determined that the average human required 6.7 **global acres** (an average land area needed to grow and produce) per person to sustain a life style with goods and services, while the sustainable global acres available is only 5.2 per person (and shrinking rapidly with the population growth). The shortage of 1.5 acres per person (29%) is made up of nonreplicable supplies like fossil fuels, and overexploiting such resources as fresh water, forests and fisheries[17]. Overexploiting means that we are depleting the supply faster than it can be replaced. In other words, fewer and fewer fish in the seas, trees on the land or oil under the land. The projection of the world population by the year 2050 is ten-billion (10,000,000,000) people. At today's average living style, that would reduce the available sustainable global acres from today's 5.2 to only 3.6 per person. Even without the anticipated increase in demand by developing countries, the shortage in sustainable land would grow to 84% from the current 29%. That means the overexploiting of our environment will need to increase at least 190% from today's situation in order to maintain living standards.

We are currently living beyond a sustainable life style in areas of fossil fuel consumption, fresh water consumption, population growth, pollution, agricultural techniques, raw material consumption and toxic waste production, and the situation is rapidly getting worse. This simple truth can't be explained away or avoided. These natural limits won't just go away. When we reach those limits, that's it. Things stop. No way around it. When we can't put fuel in a car, it does not go anywhere. When the water faucet is dry and the store is out, then what? We are already living well beyond what is sustainable, and there is no massive government effort to solve this, no international crisis team, no real action. This is not something that can be postponed until a nice political climate comes around. The political climate will only likely deteriorate when these problems stare us in the face and hurt.

To see how exploiting resources can lead to trouble we look at the oil situation. Oil is a fixed supply resource. It formed several hundred million years ago from the decay of plankton on the ocean bottom and became trapped in sandstone. Oil production has recently increased due to fracking while at great expense to the environment. The more extreme efforts at getting the last oil only delay the fact that it is running out and by mid-century is likely to be gone. Shortages of oil will happen sooner or later this century.

Overfishing of the oceans is rapidly leading to global fish depletion[18]. Studies conducted over the past fifty years on all sixty-four of the large marine ecosystems in the world, show a steady decrease in fishery yields. An extrapolation of the decreasing yields shows that a world fishery collapse is in sight, occurring at about 2050. The adverse effect on food supplies will be substantial.

Fresh water is another resource we are using beyond a sustainable rate. The International Water Management Institute has issued a warning about this[19]. The institute reported two conflicting demands on current resources. One is the need of water for domestic use by the growing world population and the other is the water needed for agriculture to feed the growing population. In favorable agricultural areas, crops can grow adequately from naturally occurring rain. But in large areas of the world, additional water is needed for irrigation.

As the demand increased for domestic water, less has been available for crops. At the current rate of decreasing agricultural irrigation water, by 2025 about 30% of the current world cereal crops (including rice, wheat, oats and corn) will be gone. The increasing difficulty in getting water has led to increasing drilling to obtain below ground water. But studies between 2002 and 2013 of the world's largest aquifers shows that most are extracting more water than is coming in, making them a diminishing source of water, just as more, about 2-billion, are depending on them[20]. The United Nations predicts that by 2025 2.4 billion people will live in regions of intense water scarcity.

The trend is clearly that higher standards of living are less sustainable. As more areas of the world are developing, the issue of sustainability is rapidly growing worse. The more we overexploit resources to make up the shortage, the more we become dependent on them and the faster they disappear. The biggest efforts in many developing countries are not to develop sustainable methods, but to mimic the unsustainable methods of the developed counties. The problem not only grows rapidly worse, it is accelerating. Few major efforts in the developed counties are underway to make significant advances in sustainability. This must change very soon to prevent widespread shortages from destabilizing the global economy and leading to widespread political unrest, followed by the collapse of the world's infrastructure.

Hunger and Poverty

In today's world, although we are living beyond a sustainable level, there is enough food to feed the world's population and enough wealth to provide a decent living for all people. The reason for starvation and poverty is, in part, the political realities and, in part, the fact that too few people are really concerned enough to make change happen. It is not just a simple matter of donating some money to help the needy, it is a complicated task. Personal and corporate greed, and a hesitation to sacrifice for others, are part of the problem. Poverty and hunger are not just unfortunate states of being for many people. They also lead to political instability, violence and terrorism that end up affecting us all. People living without hope become desperate and easily fall into the hands of those who want to use them.

Violence

In the twentieth century, there were from 125,000,000 to over 165,000,000 people killed (depending on various sources) in wars and by tyrants such at Mao Zedong, Josef Stalin, Adolph Hitler, and Pol Pot. This does not include small or individual acts of killing. These statistics justify adding violence as a major world problem, especially when we realize that, due to advances in technology, we've become even more deadly in our violence. At the same time, the world has not become proportionally more wise or restrained in its use of violence. In the US, computer games and movies that emphasize violence and killing are widespread and big business. What are they teaching?

Another aspect of violence is the growing threat of nuclear weapons. After the end of the cold war, there was a great relief in the feeling of a diminished threat of nuclear war. But today we now have nine countries with nuclear weapons and more on the horizon. New studies, made possible with advanced computers and newer sciences, have shown that the possibility of nuclear winter from a regional nuclear war is real[21]. If, for example, India and Pakistan, countries which engaged each other in several wars in the twentieth century, exchanged their current nuclear weapons, the results would have global impact. Not only would more than twenty million people die from the direct affects in both countries, but a decade long global chill would cause years without normal summer. A small example of that happened in 1816 after the eruption of Tambura in Indonesia. That eruption led to crop killing frost in each month of summer in New England and widespread crop failures. In Europe it led to a stock market collapse. That was only one year's worth of cold, not the decade that a nuclear exchange could cause. One billion people would be exposed to the serious threat of starvation, and financial systems would be exposed to great stress.

Tyranny is a more subtle form of violence. Two major kinds of non-individual tyranny and aggression are rampant in the world today. One is through the use of political power, including military and police power, the other is through corporate greed and the power of money and influence. Corporate power enslaves more people today than political power by way of controlling people's livelihood directly (through polices and the work environment) and indirectly through the political power that big money and influence wield. The results show up as the increasingly high levels of stress endemic in everyday life in the world today.

Terrorism

Terrorism is also an act of violence, but it is a special case as it is frequently unpredictable and occurs against otherwise "innocent" people. Until September 11, 2001 (commonly called "9/11") Americans did not think much about terrorism in the US. The four coordinated acts of terrorism on 9/11 changed that. Within a couple of hours, thousands died. The economic impact, still being felt, is in the billions of dollars, not to mention the emotional and social ramifications. The reality of a new era of terrorism was born in the US that day. The ability of small numbers of zealots to cause havoc in modern civilized countries was graphically demonstrated. In time, organized zealots, with the support of a zealot country or two, will gain access to chemical, biological or nuclear materials and more sophisticated methods of using them. Living in a world where massive acts of terrorism occasionally occur may be unavoidable with the current world political climate.

As long as we have rich countries and very poor countries, there will be people living without hope who become desperate enough to commit acts of terrorism. They will include in their targets the rich counties and those who they see as enemies of their ideals. While large areas of people live in hard conditions and without hope for a better life, the continued production of zealots is almost guaranteed. The cycle of hate continues to repeat endlessly without any real solution in sight.

Political Gridlock

The political atmosphere in America has deteriorated into open warfare and seething hatred with gridlock replacing a proper functioning government. The amount of negative campaigning and mudslinging has become so bad it would embarrass a pig. Compromise has become a dirty word. How can we, in such an atmosphere, move forward in any reasonable manor? It more closely resembles a 19th century battle ground with two armies facing off for a fight to the finish than a mature democracy functioning for the good of its people. Such a government has little chance of affectively addressing the global problems we now face.

Combined Effects of Issues

Each of the issues discussed above is formidable in itself. In many cases efforts to solve one problem only adds to others. One example is that in order to meet air pollution standards, some electric generating companies reduced air pollution form their coal fired electric generators, with the side effect of increasing water pollution problems[14]. Another example is synthetic fuels developed to reduce oil consumption impacted food supplies and ended up consuming even more energy than they saved. Because these issues are occurring at the same time, the magnitude of the problems becomes considerably more severe. We have to work on and solve all of these problems simultaneously and soon.

It is here the words of Albert Einstein "The significant problems we face cannot be solved at the same level of thinking we were at when we created them", take on added significance. He was saying that with the same mind sets, the same corporate game plans, the same political structures, and the same group pressures, we produce more of the same results. Fortunately, as we shall see, a new level of thinking is forming just in the nick of time.

The discussions in this chapter are brief summaries of what could be expanded into a whole book and are not complete or in-depth for any topic. Such a book could be titled "The Bad News", or even "The End is Near". However, the purpose of this book is to look at the problems from an entirely different perspective, one that brings a surprising and unexpected twist to why we are in such a mess, and with a positive and optimistic conclusion. Before we can do this we need to lay the foundations.

2

How we think affects how we live, how our society functions and how our government governs. –Robert B. Calkins

Why the Problems?

The following discussion delves into some reasons for the aforementioned problems. It is not intended to place fault or suggest solutions. Rather, it is to list some of the apparent underlying conditions in our current culture that may help produce these problems, either directly or indirectly. The reason for looking at this is to better understand the underlying type of thinking producing today's issues, the level of thinking we need to rise above in order to truly solve these problems. How to rise above them we will examine later. Remember, this is not about assigning blame, just observing what is, without judgment. While the first chapter was the bad news chapter, this one has only a modest negative flavor. The next chapter starts a change in mood which leads to a more positive attitude thereafter. Don't change channels yet.

Our Economic Model

It may not be immediately obvious that the economic model underlying our economy has much to do with many of the problems we face today, but it does. One central aspect of our model is that it is heavily based on economic growth. Things like social security, retirement plans, and debt are based on having more of everything in the future to pay the bills that will come due. Without this expansion, the numbers do not add up. There would not be enough money to pay off today's commitments. Continuous expansion has been the rule sense World War II ended, save for occasional short recessions. We have become hooked on expansion like a drug.

This expansion has been significant over the past sixty-five years and is generally expected to continue that way indefinitely. The problem comes when we realize that the world population is already living beyond long term sustainability, as noted in chapter 1. Sustainability rapidly going from bad to worse and an economy based on continuous expansion, are like two trains at high speed on the same track approaching head on. Blind desire is driving the train of sustainability and the driver of the train of Economy is not looking where it is going. There is a mathematical conflict here that is not going away with wishful thinking. Other than that, things are fine.

Continuous expansion means ever increasing consumption. People have become accustomed in the US to an ever expanding standard of living and having more of goods and services as time goes on. More people all the time want and expect more. Meanwhile, the natural resources continue to diminish at an ever increasing rate. Our current economic model is therefore itself unsustainable.

Our economy is also based on disposable products as never before. The consumption of products in the United States today is well over the combined total of more than a hundred of the least consuming countries[1]. We generate over 500,000,000,000 (500 billion) pounds of municipal trash annually in the US[2]. This does not count all the other trash that does not make it into a municipal dump.

We are constantly opening up packages (bottles, cans, plastics, boxes, paper wrappers) and throwing the container away. It is a national habit, one hard to stop. So much of what we want or need comes that way. You can't just go to the market and buy a handful of ice cream (Yuk, it would melt and end up all over things on the way home). We have begun to recycle more as a nation, which helps reduce the use of new natural resources, but the rate is modest compared to continued growth in consumption.

Technology changes so fast that a new product with improved capabilities comes out every year or two. A person with a five year old cell phone gets laughed at. A five year old computer is so out of date it is referred to as a "boat anchor". Last year's computer game is just not "with it". Last year's fashions will not do either. What, your TV does not have a gimerrod on it? (I made up that word, but by the time you read this, someone will probably be marketing it and you will need one).

There was a time when products were made to last and to be repaired. Today, planned obsolescence is common, making the cheapest possible product. Have it made somewhere where labor cost is the lowest possible, cut corners in every way and produce a disposable cheap product that will not last long. This consumes natural resources and ends up in the dump in a short time. Cheap is expensive.

Our economic model is also based on what humans have always done, take from nature and walk away. In early times of humanity, this left a small foot print on nature, one that nature could easily take care of in a short time. As we have multiplied so successfully as a species, this foot print has grown at an exponential rate, just as our population has. Our efficiency at taking from nature has massively increased, also at an exponential rate. At some point along the way, taking from nature became more the rape of nature.

The economic cost of repairing the damage done to nature is large and not yet part of our way of acting. If only a few companies did this, they would be at an enormous economic disadvantage in the market place, so it does not generally happen. As a result, actions like strip mining and clear cutting of forests are common (figure 2.1).

All this has produced a consumer mentality in the US. "Things" become important and the way they are produced or their effects on nature or our environment are often not considered. In some ways "things" become more important than people. Having our things here is not considered when people elsewhere are living in dire poverty or suffering from starvation. The two don't seem connected. Living amongst the mass consumer environment distracts people from other happenings and makes it easy to not think about other peoples' problems. The problems of the first chapter are too big to handle, so people tend to leave them to the governments to solve.

Figure 2.1. Clear cutting of a forest for lumber in Washington State.

Corporate Culture

The corporate culture in America is closely tied to the economic model. Corporations are run by **CEO**s who report to the board of directors and the stockholders. They are expected to provide a growth in profits and sales of over ten percent a year. If they don't, they are replaced. The atmosphere in corporate America is therefore aggressive. Perform or you are out. Continuing to provide profits and run an efficient and stable company is not generally an option for public traded corporations (those who have publicly owned stock). This has led to the situation where a number of large corporations seem to be run by sociopaths. The CEOs have to make economic decisions that are not in the best interest of their employees, the general public, our nation or our environment, but benefit the corporate bottom line, or their own bottom line.

It has come to the point that CEOs have made many millions of dollars a year, even when doing a disastrous job. They have sometimes been put into the place of having to make decisions that are best either for their own profit, or that of the corporation. Such a conflict of interests is not a healthy situation.

Have you wondered why all the mergers and buyouts have been occurring? When you must show double digit growth (more than 10%) there are several ways to do this. One is to increase a corporation's share of a market. But everyone else is trying this also, so it is not very easy. A corporation can also develop more product lines. But this takes time and money, both limited, and can backfire if the new product lines do not do well. A third method is to buy another company, so that when they add the number of products of both companies together it shows higher sales and looks like growth. That looks good in the annual report. The problem with this approach is that it needs to be repeated year after year to keep the growth going. The frenzy to buy other companies begins to occupy much effort, and, as time goes on, the corporation is filled with all sorts of products and splinter efforts. Big is not necessarily efficient and it is slow in reacting to change. It tends to lead to bloated hierarchies led by people who lack concern for the results of their actions. The pressure is to successfully purchase, not necessarily make the wisest choices or wait until the timing or price is right.

Another aspect of our current corporate culture is greed. Personal greed seems to be the norm. This makes successfully running of corporate America more for the executive's personal wealth and gratification than for the corporation. Executive pay started escalating about 1980. There have been examples of CEOs getting in excess of $100,000,000 in a year and retirement bonuses far in excess of that. The ratio of CEO pay in Europe, in comparison to the average corporate employee, is in the range of 10 to 1 up to 25 to 1. In the US it is approaching 500 to 1[3]. This represents a loss of balance and reason. Such people are out of touch with the rest of society. CEOs become more like kings than executives (instead of "off with their head", it's "you're fired"). This is a problem.

Excessive personal greed of CEOs does more than just affect them personally. Their actions taken out of greed are often not in the best long term interests of a corporation. Large corporations are not just curiosities; they have an impact on employees and the country in a big way. When a large corporation has major problems, it impacts thousands of employees directly, and impacts the nation through the size of its economic foot print. We have seen in the economic disasters of the 2008 to 2009 time frame how it can impact the economic health of the entire world. It is an unhealthy situation when the actions of a handful of greedy CEOs can do this. We should not blame CEOs only, however. Greed is something that has become much more pervasive generally in the US over the past several decades.

Corporate power has been a growing influence in America for many years. Today, the vast majority of our economy is controlled by well under one percent of the population. Worldwide, more than 95% of the economic power is controlled by less than 0.0001% of the population[4]. That kind of power is not healthy for the world at large. It takes only a tiny handful of people to seal the fate of billions. This small number of people are so far removed from the daily life and struggles of the masses that they are not impacted personally by many of the things most people are. This privileged few live an isolated life where the decisions they make have an effect on billions of people but are not necessarily affected themselves. This is done out of public sight and often without any public knowledge.

The idea of democracy is that the people elect the leaders of government to represent them in running the country. Corporate power is controlled by CEO's who are appointed by the board of directors, who are only elected by stockholders. Corporate power has become so big it rivals government with the effect that democracy has been compromised. When corporate power has a major impact on whom is elected, on lobbyists, Congress and government agencies, it becomes in effect more like a plutocracy (rule by the wealthy).

Government agencies that regulate commerce are heavily influenced and partially controlled by the corporations they are regulating. Large corporations spend huge amounts of money to make sure they have the influence needed to protect their interest and to influence elected officials, corporate regulators, voters, and even the news.

You may ask what does all this discussion about corporate power have to do with the world problems of chapter one? These problems are too big for one government; they need a cooperative effort of many governments. If these governments are controlled or partially controlled by large corporations, and the interest of the corporations come first, then the interest of the world's people come second. We have seen how the monetary interests of large corporations have stymied effective efforts to tackle global warming. The same is true for the other problems.

The simple facts show that the pressure to provide stockholders with significant growth each year, combined with the extreme pay levels of executives who are trying to keep their jobs, corporate interests become focused on ways to stave off anything that costs money or puts restrictions on them. The corporations, using the expertise of Madison Avenue advertising, can and have kept the public confused about what is real and have put pressure on the government to hold off or downplay regulatory action.

This situation makes it hard for governments to be bold and do the unpopular things needed to really solve these problems. The level of effort needed internationally to make real progress needs heavy popular support before the politicians are going to risk taking actions that are sure to hurt the economy. That has not happened yet. Most of the world's population are too occupied with earning a living and struggling with daily life. They are not yet raising the issues to government.

Loss of Community

The sense of community of past generations has largely been lost in much of the US. There is a meaningful psychological value in the sense of community. It includes the perception of being among others like ourselves, of a degree of interdependence and of being part of a larger and stable structure. This brings a feeling of belonging, meaning and safety. It also leads to a greater sense of civic involvement and awareness.

Being a part of a community improves familiarity and identity. Knowing shop owners and service people, local history and where to find things are all a part of community. There is a greater chance people will participate in a community they feel a part of.

In our modern society, mobility is more common now than community. Email, text messaging, computer dating and fast food outlets, each widely common, replace community and real in-person contact. (One result is people eating unhealthy food alone in front of TV.)

When people feel less a part of their community, they are less apt to be aware of local problems or be involved in finding solutions. This can lead to a sense of detachment about bigger problems.

Ancient Religious Customs

In ancient times, organized religions provided a needed and important function for humanity. Most people were illiterate, ignorant and lived superstition filled lives under harsh conditions. Violence was common. Religions provided a stabilizing influence on regions and in many cases were, in effect, the government. Ignorant peasants had a sense of belonging, order and hope not otherwise present. This was an important step in civilization.

Some would point to the violence that occurred through organized religions or under religious banners. That should be put into historic perspective. Those were relatively primitive times and widespread violence would have been occurring without the religions, probably more so.

In the 21st century, large portions of the world's cultures are dominated by ancient religious customs and beliefs that stem from times hundreds or thousands of years in the past. When many of these customs were formed, few people could read, most people were peasants, slaves or serfs. The world was perceived as flat and covered by a large dome, magic and superstitions were widely believed in and very few people knew anything of the world beyond where they had been born. Yet some of the customs developed in those primitive times are held as ultimate truth even in the face of modern science. When customs over thousands of years old will not change in the light of greater understanding and awareness, progress and problem solving can be severely hampered.

One example is the religious taboo against any form of birth control. At a time of population explosion and living beyond a sustainable consumption rate, efforts to control the size of the population are important. When ancient religious customs dig in their heels and refuse to change, the problems are much harder to solve. This is especially hard when a belief is held that God will save the "chosen" ones at the end of the world. When many believe they are the chosen, why worry about things now?

Another problem that ancient customs present is an attitude toward women. Keeping women in subservient roles is wasting half the human creativity by ignoring it. The belief by male dominated religions that woman are inferior is a tremendously wasteful belief, not to mention degrading.

The belief that God provided nature for man's pleasure and use is also a limitation. The attitude is "there is nothing wrong in exploiting it to the fullest". After all, God gave nature and the Earth to us. There is nothing in the old religious texts that says we have to limit our use of it or that we are responsible for intelligent management of it.

The ancient traditional ideas of "our religion being right and others being wrong" have led to massive violence and hatred over the centuries. There is fundamentally a problem with religions that claim to worship a loving God, while preaching hate. Preaching "we are right and they are wrong" inevitably leads to conflict, hate and violence. When "we are right and they are wrong", how can we work together to solve worldwide problems?

A lot of good has been done by religions throughout the centuries. Many loving and sincere people have been involved. Yet, as progress is being made, it is the ancient establishment that changes slowly and seems to be holding progress back.

Old Habits

Well established cultural and sociological habits are hard to change. People, organizations, companies, governments and cultures at all levels tend to continue doing what they have been doing by habit. Habits alter our perception of reality. They give us a feeling things are all right when they may not be. The larger the number of people participating in a habit, the greater the inertia and the harder it is to change. In rapidly changing times like today, conditions tend to change faster than habits. Problems can get worse even while action is being taken to compensate. With this in mind let's look at some of our cultural habits that make solving our problems difficult.

The very first humans were hunter-gathers. They lived by taking what they could use from nature. They had little impact on the earth because of their small numbers. However, the deep seated habit of just taking and using nature has continued up to today.

Technology has given humanity such power that it is no longer true we can just take and use nature without having a worldwide impact. Changing our habits to become a long term sustainable culture is hard, it takes farsightedness and substantial costs.

One habit in many parts of the world is fishing for a living, either for food or as a business. Many generations of fishers have produced a habit that has increased in strength as the population has grown. The eating habits of many continue the demand for more fish. Trying to change deep seated eating and job habits is hard and strongly resisted. Yet without changes, the seas' fish supply will soon be exhausted.

Changing these habits is politically difficult. It takes a real sense of urgency before the level of change needed is likely to spur governments to act. Even if individual governments act, it may not solve the problem if others take advantage to continue old destructive habits. An example of this is the attempts to save the whales, undercut by the Japanese who continue to conduct whaling. The Japanese are one of the most fish eating people in the world and it is much more difficult for them to change centuries of habit just to save the whales.

Another centuries old habit is dumping of wastes. For a long time manufacturing has just dumped waste products wherever it was most convenient, in pits, rivers, oceans, etc., leaving once-natural sites in unusable condition. Individual people have done much the same.

A form of habit comes from the fact that there have always been trees, clean air and water, fertile farm land, plenty of fish in the oceans and rivers, and plenty of minerals. We have a habit of thinking of these as being there and available. It is therefore difficult to really believe this is no longer the case. We can still get fish and fresh food at the market, there is wood in the lumber yard, potable water comes out of the tap and there is gas at the gas station. Can it really be all that bad? The answer is no, not at the moment. But if major action is not taken soon, this will change. Again, it is our habitual way of seeing reality that hinders the full appreciation of the problems.

Another topic of habit is freely reproducing. Reproduction, as a basic biological drive, exerts a strong influence on people and culture. In many parts of the world, producing a large family is a basic and habitual part of culture. In some ways it is a form of social security for parents, having enough children so some survive and care for the aging parents. But with the population explosion being one big aspect of our problems, changing this habit is necessary to bring the population growth to near zero. The depth of cultural change for this is massive in some parts of the world. China reduced its population growth rate with a one child policy. But it is a totalitarian government that is not responsive to its people. For a country like India to do this would be much more difficult.

All of these world problems are bigger than one government. They are global problems and it will take a concerted effort by the world governments to solve them. This in itself takes a change in habit. For hundreds of years, national governments have been making policy for their country. International agreements have been difficult to produce and usually of limited scope. The limited success of the United Nations demonstrates this. For all the problems facing humanity today, a new habit of international cooperation is a must.

Other Causes

Privileged class pressures on our politicians have resulted in political gridlock, with the establishment running things for themselves behind the scenes. The public has become thoroughly disgusted with it all. Government of the people, by the people, for the people has degenerated into government of the privileged, by the privileged, for the privileged, producing a growing dissatisfaction within much of the public.

We have examined some, but certainly not all, of the current cultural conditions (world views and values) that have helped establish today's set of major problems. We could go on for quite a while analyzing causes or underlying reasons for these problems, but we have gone far enough for our purpose here. The bottom line is that our current culture has resulted in some very serious problems and is currently not able to handle them. Something must change and change soon. This book is about the promise of that change for the better. But before we talk about change, we have one more aspect to briefly cover in chapter 3, which is how we came to our current predominant culture.

3

Study the past if you would define the future. — *Confucius*

Origins of our Current Culture

In order to better understand some of the underlying causes of today's world situations, it helps to look at the basic cultural thinking patterns and world views that drive society and shape our reality. We have examined some of the serious problems of our time and some of the reasons for them. Now for a brief look at where our current world views came from to better understand them and how they came about.

Humanity's past can be divided into two broad categories: pre-history and historical. Of the historical times, historians divide history into three broad periods: classical (also called "antiquity"), medieval (also called "middle") and modern.

Although writing first developed about 3000 **BCE**, the classical era in the West started about the 8th century BCE and was centered around the Mediterranean Sea (especially the Greek and Roman Empires). It lasted until the fall of the Roman Empire about 500 **CE**. During this time period, the first of the Greek philosophers emerged along with systematic observations of nature and logical deductions[1]. Powerful city states and empires provided safe and stable environments for trade and the arts to flourish. Much of the religious views were full of many gods.

The Medieval Age

The medieval era was the period from 500 CE to 1500 CE, starting with the fall of the Roman Empire. This era saw a general trend of deterioration from the days of the empire. Urbanization and population decreased. Without the safety of an empire and its troops, trading and commerce decreased due to the dangers of travel.

It was a time of great famines and plagues. Medieval Britain had ninety-five famines during this time. The Black Death killed as much as a third of the European population[2].

The majority of people were slaves, serfs or peasants. Very few people had any education. Illiteracy was the norm. Most people lived their whole life and died within a few miles of their birthplace, knowing almost nothing of the rest of the world. People were superstitious. The church controlled their lives, telling them what they should do and how to live. The only science was practiced by priests with the assumptions of God being in control of everything. What little science was conducted had to conform to the accepted religious views of the church. If it did not, those conducting the science could be arrested, imprisoned, tortured, burned at the stake or all of the above. It was a time when the world was not only the center of the universe but *was* the universe. The sun, moon and stars were on a giant dome over the flat earth. Things happened the way they did because God made them that way.

The "world view" of this time period (sometimes called "Traditional") included the ideas of sacrifice of self for the group, loyalty to the rulers, salvation through obedience to the church, and a black and white view of right and wrong.

In the later medieval age, the culture started to slowly move toward what would become the modern age. Universities developed in some major European cities, making literacy more accessible. Translations of some of the classical Greek philosophers become available, sparking a new attitude and enthusiasm for philosophy. The stage was set for a new world view. The transition from medieval to modernity lasted several hundred years, with the modern era starting in the 1500's, the time of Copernicus.

The Emergence of Modern Science

The roots of **modern science** cover centuries of ideas. The major changes in its emergence generally are recognized as occurring between 1543 and the 1700's. The scientific revolution continued through the 20th century.

When modern science emerged out of religion, the priests who did much of the early work were limited by their religious beliefs and assumptions. However, all scientists throughout time have had a version of this limit. Everyone has a set of beliefs they have to work through.

The early pioneers who brought us into modern science built on the knowledge of those who came before. Some of the ideas the great names of that period used came from the classical Greek philosophers[3].

Those Greek philosophical ideas had been lost to Western Europe during the medieval era because they were recorded on papyrus. If not re-copied at regular time periods, the papyrus fell apart and the writings were lost. When the Greek empire ended, there was no longer organized upkeep of the records. Fortunately, some had, however, been saved by the Islamic Arab Empire in Eastern Europe which collected and translated them into Arabic. Later, translations of the Greek philosophers into Latin became available to Western Europe. The rediscovered philosophies fueled a new period of advancement that led to the emergence of modern science[3]. Many people were part of this emergence. For the sake of simplicity, we will limit our discussion to six representative pioneers that illustrate the course of this development.

In the 1500's, superstition, religious dogma and fear were widespread. The world was seen as a hierarchy having been created by and ruled by God. Everything that happened was perceived as part of this hierarchy, including the rights and powers of religious leaders and kings. Ordinary humans were below the church and rulers, but had dominion over plants and animals. Men had dominion over women and children. What happened did so by God's plan.

In most of Europe, the Roman Catholic Church dominated thinking and dictated what was right and proper. The earth was seen as the center of the universe (geocentric). It was surrounded by spheres of heaven where the sun, moon and stars were affixed. The movement of these spheres caused the heavenly bodies to move in the sky. In such a world, things were basically static. Reality was fixed and nothing changed except by God's action. It was dangerous to introduce new ideas. Many people faced the inquisitions and more than a few were burned at the stake. It was often necessary to get the church's permission to publish a book. For science to emerge, it had to move beyond the domination of the church and its concepts of hierarchy. This was a difficult separation.

Nicolaus Copernicus (1473-1543) lived most of his life in Prussia. As with many early pioneers, he had diverse interests and was a mathematician, astronomer, physician, classical scholar, Catholic cleric, governor, diplomat and economist. He was instrumental in advancing the idea of a sun centered universe (heliocentric)[4]. This placed the turning of the earth on its axis and rotation around the sun as the cause of the apparent motion of heavenly bodies. Copernicus had in effect removed the Earth, and with it, man, from the center of the universe and God's creation, and made it just one of the planets.

Such an idea was condemned by church authorities as sacrilege for the next hundred years. Copernicus avoided trouble with the church by not publishing anything until the year 1543, twenty-nine years after he first recorded his ideas. That year, he published his two books and promptly died, thus avoiding the inquisition that Galileo faced 90 years later. The publishing of Copernicus' book *On the Revolutions of the Celestial Spheres* in 1543 is the reason some consider that date the starting point of the emergence of modern science. Much more was of course needed to flesh out this era.

Sir Francis Bacon (1561-1626) was an English philosopher and statesman. In 1620 he published the book

Novum Organum which established and popularized the deductive methodology of science, the modern scientific method of enquiry, for which he is best known[5]. At that time, such methods had been considered a part of the occult or alchemy. Bacon insisted on using planned procedures for scientific investigation, which introduced the idea of methodology into science. He did not have to worry about persecution from the Catholic Church as England had separated itself into the Church of England years before.

Galileo Galilei (1564-1642) was an Italian philosopher, mathematician, astronomer, and physicist[4]. He is recognized as the first to use the experimental method in systematic studies of accelerated motion and astronomical observations, and to have supported the heliocentric concept of Copernicus. Galileo is regarded by some as the father of modern astronomy, modern physics and modern science.

Galileo, who lived in Italy, was tried by the Catholic Church for heresy in 1633 for his support of a heliocentric universe. The church held such a view absurd and heretical because it was said to be contrary to Holy Scriptures. He was found guilty and imprisoned for life. His sentence was later changed to house arrest. He lived the rest of his life confined to his home, prohibited from communicating with others. During those days of solitude, he continued his work which culminated in one of his best books, which later received high praise from both Albert Einstein and Sir Isaac Newton.

Johannes Kepler (1571-1630) was a German mathematician, astronomer and astrologer (in those days astronomy and astrology were part of the same discipline)[6]. Kepler had started out to become a minister, but ended up a teacher instead. He discovered that the planets traveled in ellipses and showed how to calculate their paths. Before then, it was believed they traveled in circular orbits. Kepler developed laws on orbital mechanics that were the foundation for Newton's law of gravity. He also did important work on optics and helped validate Galileo's findings.

Rene Descartes (1596-1650) a French philosopher and mathematician[7] has been called the "Founder of Modern

Philosophy" and the "Father of Modern Mathematics". Descartes invented the Cartesian coordinate system resulting in graphs and graph paper. He also invented analytic geometry and writing numbers with superscript (like "x^2" for "x squared"). He is famous for his quote "I think, therefore I am".

Descartes developed the idea of radical **dualism**, where the body works like a machine following the laws of nature, while the mind is nonmaterial and does not follow the laws of physics. This separation of the mind from objective nature had major implications for emerging modern science. For Descartes, reason along with direct observation, was necessary to understand the workings of nature. Purpose was seen as an exclusive property of mind and not the physical world. This view would become a backbone of modern science for the three centuries to follow.

Descartes' first book was intended to be published in 1633, but the trial of Galileo scared him off (He did publish his book a few years later). In 1663, the Pope placed all of Descartes' books on the prohibited book list, but Descartes had passed away by then.

Sir Isaac Newton (1643-1727) an English physicist, philosopher, mathematician, astronomer and alchemist is considered by many to be the single greatest person in the history of science[8]. He published his ground breaking book *Philosophiae Naturalis Principia Mathematica* in 1687, which presented his ideas on gravity and his three laws of motion. To this day, these are taught in all basic physic classes. Newton co-invented calculus along with Gottfried Leibniz.

The results of Newton's laws were that the natural world was seen to behave like a machine, providing the theory that supported Descartes ideas. These laws supported the growing belief in **materialism** and **determinism** that became the core ideas of science for several centuries.

Along with numerous other contributors, the ground work for modern science was firmly established. The process of separating science from religious dogma, fear and repression was mostly completed in the 250 year time period covered above, although struggle continues.

The Decline of Religion

In medieval days of the West, the church dominated the culture and lives of most people. Science did not exist as we know it and what little was practiced was unknown to the vast majority of people. Life was simple.

Religious dogma built up to provide guidance to the masses and to establish control and authority of the hierarchal male dominated organizations. The ego of the religious rulers also played a part in dogma. No one could safely challenge it.

As the transition to modern times occurred, science slowly struggled to find its place free from the domination of religions. The more science was successful at this, the more scientific findings challenged the established dogma. Religions often resisted changing dogma in the face of scientific evidence, thus keeping science and religion in conflict, something still occurring.

In the 20th century, science made spectacular strides, resulting in an onslaught of new technology. This success, in the face of the resistance of the old religious hierarchies, has steadily eroded religions credibility. Strong national governments and countries have eroded the need for and authority of the religions that dominated Western culture in medieval days. Male supremacy and domination in most religions are less and less acceptable in the modern Western society. The old anthropomorphic image of God, still evident in some religious teachings, is also viewed by many as out of date.

Religious texts written thousands of years ago in primitive times by people who did not understand reality the way we do today determined religious dogma and custom. To many modern people, these are seen as behind the times and archaic. The texts refer to cultural idioms, slang and customs so far gone that only a few experts can understand, and few ministers are such experts.

A well-educated, scientifically and technologically aware mobile public has much more to fill their lives than people did in past eras. Religions have declined, in part because their dogma contained things that were not in agreement with modern times. Inaccuracies became obvious to an increasingly educated public. As the religions resisted admitting their errors, they lost credibility while science passed them by.

A global comprehensive study of religious trends conducted in 2009 shows an increasing number of people with no formal religious affiliation, and the strongest participation coming from the older population[9]. In the US, 65% of those in the "Millennial Generation (Ages 18 to 29) rarely or never attend worship services[10]. Between 1958 and 2000 in the United States, church attendance declined from 49% to 40% of the population[11]. In Australia from 1961 to 2000, church attendance shrank from 40% to 24%, with a similar result in New Zealand. In Britain during the same period, attendance diminished from 18% to a miniscule 7.5%, raising speculation on the remaining viability of the Church of England[12]. A study in Spain that covered the time span 1930 to 1992 showed an average annual church attendance drop of 2% per year[13]. Figure 3.1 shows the changing attendance and the average age of attendees in the UK[14].

Similar trends have occurred in Canada. A 2009 report on the United Church of Canada (UCC) offers a dim future[15]. Using data going back to 1945 it examines the trends and projects ahead to 2025. In 1945 the UCC membership represented 6% of the Canadian population. In 2007 that had dropped to 1.6% and projected to 2025 at the current rate it would be 0.4%. Membership went from its high of just over a million to just over half that number in 2007, and could drop by a half more in the next 18 years. The number of churches has gone from 6780 down to 3420 and is projected to be 2490 by 2025, with smaller average attendance. UCC school membership is dropping so fast that it is projected to reach zero in a few years.

The overall story of churches is one of decline. Fewer churches, smaller and rapidly aging congregations with declining revenue, fewer ministers with fewer schools and seminaries are in evidence. In some cases, the question is not about the health of the organization but about when to close it. Organized religion however, is not going away or on the verge of dying out worldwide. Mainly, its cultural significance is decreasing in the more developed countries.

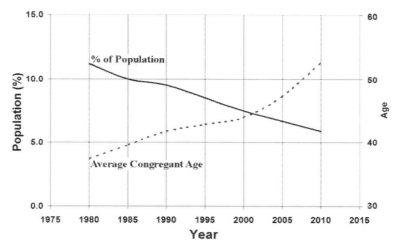

Figure 3.1. Church Attendance in the UK

The Modern Age

The modern age has continued to recent times. Our current major cultural world view in the west is called the "**modernity**" (not to be confused with **modernism**, a more recent quality). It grew out of the medieval age over several hundred years. Modernity was sparked in response to the problems and limitations that the medieval age created. This does not mean that modernity is the ultimate world view. It also has created problems due to its limitations. This is a good place to look at the general characteristics of modernity.

Along with the rise of science came the ideas of the world being open to human intervention in place of the idea that God ran everything. Complex nation-states, industries, economies, democracy and personal freedom along with competition and striving for the good life were introduced in the modern age. The down side of the modern world view is that there is a tendency to be exploitive, unscrupulous, selfish, greedy and materialistic. So it's not perfect – all cultural periods have had a good side and a down side. Which side dominates makes a big difference and, over time that balance tends to fluctuate.

The United States of America was founded on modernity ideas. "We hold these truths to be self-evident…" was revolutionary in the history of mankind. A government of the people, by the people and for the people! How outrageous. In a way, the opening lines of the American colonies' declaration of independence and the US constitution can be looked upon as a statement of modernity. They marked a distinct and massive break with history. The words were a bold, audacious statement that things were now, and going to be, different. And they were.

The Struggle Continues

The struggle between traditional and modern views did not end with the rise to dominance of the modernity culture. People who still follow the traditional ideas, mainly around religious beliefs, believe they are right and point to the weaknesses of the modern views. There has always been a fanatical group (fundamentalist) within the traditional camp that spends considerable effort trying to force its views on the larger society. This is based on the "holier-than-thou", "I am right and you are wrong", "I'm going to heaven and you are going to hell" attitude. This way of seeing the world shows up in most religions. Many people are perfectly happy with a live-and-let-live philosophy, but extremists insist others live by their standards and campaign to force the rest of the population to do so, figure 3.2.

Origins of our Current Culture

Figure 3.2. Resistance to change.

In the US, struggles occur almost daily in school boards, local government meetings, political campaigns, libraries, with advertisers and the media. Pressure from fundamentalists have rewritten school text books, banned books from libraries, scared advertisers from supporting TV programs, closed web sites, pushed religiously based laws though legislatures and much more.

Now that we have taken a look at where our current world views came from, we switch subjects and look at some of the newer fields of science that have come along in the past several decades. The reason for this consideration will become clear in later chapters as the story of our times comes together.

Part II

A Scientific View

4

The only difference between a problem and a solution is that people understand the solution. –Dorothea Brande

Recent Developments in Science

During the twentieth century, the world went through the most amazing changes in history, perhaps as much as in all the rest of history put together. The changes were so large it's hard to appreciate the magnitude. No one alive today really remembers the times at the start of the last century (1901) and we are so far removed from that life style today that we are not even aware of most of it. Imagine the time without electricity, dish or clothes washers, or refrigerators. This was a time before automobiles, aircraft, general use of electric lights, radio or television, telephones, motion pictures (especially in wide screen, color and with sound no less), and paved roads. Think about the change from before the first airplane to driving around on the Moon, space stations, photos from all the planets, and weather reports from Mars to photos of your house from space on your cell phone, or even reading this on a hand held electronic device. Think about the change from one mile per hour horse and buggies on dirt (or mud) roads to 70 mph cars (with music, air conditioning, GPS, cruise control and soon, self-driving) on super highways. A lot has changed. The twentieth century has been called the revolutionary century and nowhere has the change been more striking than in science.

At the start of the twentieth century, science anticipated realizing a long held dream of becoming the master of nature and being able to mold it to the will of humanity. One scientist expressed pity for students who would follow him on the path of science because they would have nothing more to do than to measure and observe things in finer and finer detail. All the discoveries would have been made[1].

New discoveries soon pushed this dream further and further away. In the last three decades of the century alone, discoveries came so fast it was hard to keep up. Most scientists became increasingly specialized just to be able to stay current in their field. Many new scientific fields emerged. Some of these, taken together, weave a fascinating new perspective on the current world situation. This chapter discusses several of these newer science fields briefly, enough for an overall idea of what they are about. They will be referred to later, as the story unfolds.

In addition to science, several technologies (products of science) play an important part. The computer and the internet are two that together are reshaping the world. For example, in 1974, 15 years **BI**, (Before the Internet) most of the world's populations were unaware of cultures other than their own. Individuals were born, grew up and died knowing only their local **culture** (a condition that was not so usual in the US). Today, the vast majority of the world's population knows about other cultures. Information travels so fast, it is common for a popular song to be heard in just about any part of the world while it is still popular in its place of origin. Being aware of other cultures has a significant effect of expanding one's perspective. Something that might have been "just the way things are" in one culture now stands out as one of many options. Other ideas abound. The scope of social change that this fosters is difficult to contemplate. All one has to do to radically change an old culture is to introduce the computer and internet.

An example of the spread of information by the internet is when my daughter was traveling in China. She took digital photos of what she was doing, including the food she was enjoying. In the evening, she would turn on her laptop computer and load the better photos onto **Facebook** (an internet site[2]). Later that day, my wife and I would see where she had been and what she had been eating hours earlier. An idea can go worldwide in minutes.

The following are snapshot descriptions of some scientific disciplines. While they may not seem particularly important at this point, they will later.

System Theory

The idea of general system theory was first published after World War II, although it had earlier roots[3]. It began to make major advances in the 1960s and matured in the 1970s. It has many special applications such as in cybernetics, topology, information theory, game theory, complex **adaptive systems**, system dynamics, systems engineering and living systems. System theory takes a big picture look at a system, how it interacts with subsystems, and how it relates with the elements of the system and its environment.

System theory is an important tool in order to understand and manage complex systems of all types. It added "process" to the understanding of system function. It introduced the idea that a system works by processes and understanding the processes is as important as understanding the properties of its individual parts.

System thinking had a substantial effect on the western world view. Prior to it, science had the mind set of thinking that the characteristics of an object could be understood by examining the properties of its parts (the reductionist approach stemming from René Descartes in the 1600's)[4]. A living system is indeed more than just a pile of parts. It's an integrated complex system with properties not found in any of its parts.

The qualities of a system arise from the organizing relationships of its parts. A human being (a complex system) can get up from a chair, cross a room, leave a building, cross a street, get into a car and drive away. No part of or collection of parts of a human can do this (thank goodness).

Studying the parts of a human will not lead one to understand fully how this can occur.

One example of the limitation of studying the parts to understand the whole is contained in the story of a group of blind people that come upon an **elephant**. None of them knew what an elephant was and each felt a different part; a leg, tail, ear, belly, trunk and a tusk. Each described the part they had encountered. They debate the nature of what it is, each from their own perspective, while none comes close to understanding what an elephant really is. Even if each blind person felt each part, they would still not understand the whole.

An emerging system theory category is living systems[5]. Newer understanding of evolution has found that there is much more to it than the chance variation (later called mutations) and natural selection of Charles Darwin's day (1850s). Living systems demonstrate ever-increasing diversity, complexity and system cooperation. Life is constantly reaching out to find new avenues of diverse expression.

Another aspect of system theory is dynamical systems theory. Calculus was invented independently by both Sir Isaac Newton and Gottfried Wilhelm Leibniz in the 1600s. It gave science a powerful new mathematical tool to work with and allowed many physical phenomena to be described by mathematical equations. However, many of these equations were too complex to be solved. That problem led to a common practice of making simplifying assumptions to reduce the equations to solvable size. That was a very effective approach, but it often resulted in missing some of what was really happening.

With the advent of practical computer power, a new era blossomed in the 1960s, and with it the ability to solve previously unsolvable non-linear equations. This is what "dynamical systems theory" is about. The world of nature is mostly non-linear and to understand it fully, one needs to solve non-linear equations. Many new ideas started flowing out the mid-1960s as a result.

Self-Organizing Systems

Self-organizing systems, also called "adaptive systems", are systems that resist the tendency of **entropy** (the slow decay due to energy loss from friction or heat loss within a dynamic system). They maintain the system structure within certain limits through interchange with their environment. Living systems such as bacteria, cells, people and even societies are self-organizing adaptive systems. A common characteristic of a self-organizing system is that it is a network pattern, that is to say, the components of the system are arranged in a network[8]. Networks are non-linear communication systems. These networks also operate as feedback loops, the critical function that enables a system to self-organize. When the system does something, it gets feedback about the results and can then modify its actions.

A simple example of a feedback loop is a thermostat that regulates the temperature of a room with an air conditioning system. When the room temperature drops below a set value, the thermostat sends a signal to provide heat. When the room heats up to a set amount, the thermostat sends the signal to stop. Self-organizing systems take in energy from their environment and give off heat (entropy) and waste. They are termed an **open system** (as opposed to a closed system that does not interact with its environment).

Living systems are complex self-organizing open systems that contain a network with feedback loops. In such complex systems, tension is necessary and inherent in keeping and maintaining the system structure[9]. Social systems fall into this category and tend to maintain the status quo (a state of equilibrium or stability), which is the self-organizing system maintaining its structure.

Chaos Theory

In 1961, Edward Lorenz, a meteorological scientist, was working with an early computer to model weather in order to better forecast it[5]. After a run of the weather simulation model, which took considerable time with the slow early computer, he decided to rerun it to check on an additional item. To avoid the long run time, Lorenz took the variables from the printout at a point part way through and started the computer from there. He expected the same results, but when he looked at the data, it was considerably different. After checking all his inputs and verifying they were the same, he was stymied. Then he realized that the number in the printout did not have all the decimal points the computer had (such as a printout of 142.238 and in the computer it was 142.23786, a difference of only 0.00009%). It was so minor a difference it should not have mattered. To his great surprise, it did.

This was a major problem for several reasons. First, if extremely small changes in the input variables could make major changes in the output, how could anyone expect to reliably predict weather from a computer model for more than a short time period? Another problem was that the prevailing deterministic view of science seemed incompatible with such behavior. That deterministic view had dominated science for several hundred years, going back to Sir Isaac Newton. It held that if you knew the conditions at one point, you could determine what was to follow and what had come before. Lorenz went on to study this effect, publishing his results in a scientific journal in 1963[6]. This became the start of **chaos theory** and has since become one of the papers most often cited in other scientific papers[7].

The common misconception of chaos is that it is totally unpredictable randomness. The chaos in chaos theory turns out to be orderly chaos, unpredictable in its details, but in an orderly kind of way. When variables are plotted, they have a common area they travel in. They do not go beyond certain limits. The motion tends to be around an area, or several areas on the plot. These areas have been named "**strange attractors**". To get a mental idea of this, think about an ice skater alone on an oval rink. He/she skates around the center of the rink, the strange attractor, in similar but not identical paths that are not entirely predictable. To image a two attractor situation, as shown in Figure 4.1, think of the skater skating around on one end of the rink for a time, then crossing over in a figure eight to the other end of the rink for a while, randomly changing back and forth.

Subjects where chaos theory applies include electrical circuits, lasers, oscillating chemical reactions, fluid dynamics, mechanical devices, dynamics of planetary satellites, weather and climate. All chaotic systems have the common quality that very small differences in the starting condition produce large differences in later conditions. Systems described by linear equations are never chaotic, only those with non-linear equations. Chaos shows why even with advanced computer weather prediction models, weather prediction is not always accurate, especially for more than several days in advance.

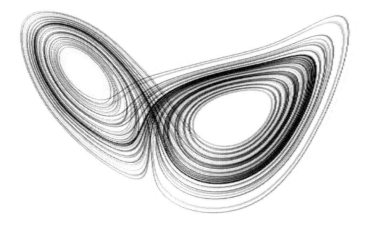

Figure 4.1. Plot of the Lorenz Attractor

Dissipative Structures Theory

The 1977 Nobel Prize in Chemistry went to Ilya Prigogine for solving a paradox in thermodynamics that was over one-hundred years old. Prigogine's work led to a number of fascinating findings. Once his idea was presented, it seems fairly simple. However, it took a shift of thinking out of the old box and the use of new mathematic tools to get there. The ramifications were significant[10]. The second law of thermodynamics, which introduced entropy, stated that all systems should move toward lower states of energy due to the losses that naturally occur. Living systems however grow in complexity, which was the paradox. Prigogine recognized that there was a difference between open systems and closed systems. This led to the solution of the paradox. He discovered a phenomenon he called "**dissipative structures**".

Dissipative structures are self-organizing systems, yet are far from an equilibrium condition (this means they are active and interacting with their environment, not static). They include all living organisms and social systems. Dissipative structures interact with their surroundings in a way in which environmental instabilities can lead them to transform themselves into new structures of increased complexity to contend with the instabilities. This transformation happens through the networks and feedback loops within the structures.

Dissipative structures can only be fully understood by studying their processes, as well as their structures and networks. They are non-linear and have an interesting way of behaving. They can handle small to medium stresses from their environment, and maintain the existing structure. But if the stress reaches an amount that the structure can no longer handle, a threshold point Prigogine named a "**bifurcation point**", a change in the structure, occurs.

Here bifurcate means "to divide into two" branches[11]. Bifurcation points are where evolution occurs because the structure, complexity and capabilities of the system are different after the reorganization. New forms or order exist after the reorganization is completed.

Dissipative structures can maintain stable forms for long periods of time until driven past the threshold to a bifurcation point by unstable environmental conditions. When the bifurcation point is reached, which path the system takes next is not predictable. The system enters an unstable period of chaos in which one of two paths is followed. See Figure 4.2. One path leads to the breakdown of the structure, the other to reorganization to a higher order of complexity and a higher threshold point; death or evolution[12]. The path taken can hinge on the smallest of factors, for it is in a condition of chaos. If the path of higher complexity is taken, the new structure enters a new period of stability. Evolution has occurred.

To illustrate the bifurcation point idea think of a stream bed. It can handle the water that normally flows in it. Every now and then a flood occurs and the stream overflows its banks. But if a record flood occurs, a state of chaos is reached in which a normal stream channel does not provide enough room for the water flow. Obstacles are encountered (like a bridge) to the greater flow and pressure builds up until the stream changes course or breaks through the obstacles (washes out the bridge). In this example, a bifurcation point is reached when the environment (flood) reaches a threshold that the old structure of the steam bed cannot support. The flood produces chaos until a change occurs that allows for a greater threshold (more water flow).

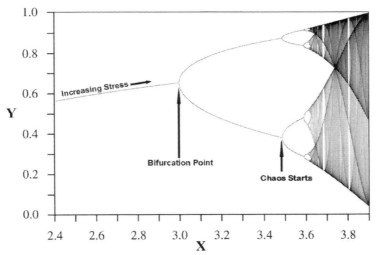

Figure 4.2. Increasing Environmental Stress leads to a bifurcation point and chaos.

Cultural Development Theory

Cultural Development theory is not simply the details of behaviors of groups of people in different parts of the world. It is about shared cultural values and cultural world views and their development. It arose from a number of different disciplines including developmental psychology, cognitive psychology, social sciences, paleontology and archaeology, as well as system theory. Cultural theory is the study of the evolution of culture, human values and world views over time. It is an important part of the big picture unfolding in this book.

Unlike most other psychologists who dealt with "sick" people, Abraham Maslow looked at healthy people and the best that people could be. He developed a list of human requirements, the "hierarchy of needs", that is often quoted today[13]. The term "hierarchy" can be misleading, so in this book I will refer to it as "Stages of Needs". Maslow's Stages of Needs, looks at development from the earliest to the latest and is: Physiological, Safety, Belonging and Love, Esteem and finally Self-actualization. See Figure 4.3.

The idea of hierarchy can be offensive to some. What is discussed in this book is not the socially imposed type hierarchy we have seen in society, with its included judgments. Here we are talking about natural stages of development that exist in nature, without judgment or social values. An example of natural stages is that of the material world. It starts with subatomic particles, and progresses through increasing complexity to atomic particles, atoms, molecules, organisms, and animals.

What Maslow found was that people are focused on the earliest unfulfilled need and don't make progress on later needs until each earlier need is satisfied. The physiological need is for basic survival and the need for things like air, water and food. The safety need is for a safe and predictable environment. Belonging and love are social needs for friendship, intimacy and a supportive community. Esteem is the need for self-esteem and self-respect. Self-actualized people have met the need to be all they can be and have reached a stage of understanding and harmony. The Stages of Needs was the foundation for further insights on human culture and its evolution through time.

Qualities	Breathing, Food, Water, Sex, Sleep, Homeostasis, Excretion	Security of Body, Employment, Resources, Morality, Family, Health, Property	Friendship, Family, Sexual Intimacy	Self-esteem, Confidence, Achievement, Respect of others, Respect by others	Morality, Creativity, Spontaneity, Problem solving, Lack of prejudice, Acceptance of facts
Need	Physiological	Safety	Loving/Belonging	Esteem	Self-actualization
Stage	1	2	3	4	5

Increasing Stages of Development ⟶

Figure 4.3. Maslow's Stages of Needs[14].

Developmental psychologist Clare W. Graves, through his studies and the work of others, including Maslow and Robert Kegan, demonstrated the developmental stages of cultural development through which individual people and cultures pass[15]. Each of these stages is a system of values, world views and beliefs unique to its stage that determines how people in that stage view their environment and react to life situations. Graves' colleague, Don Beck, expanded the ideas by noting a cyclical nature of the stages, as detailed in the book *Spiral Dynamics*. He coined the term "**MEME**" for these stages[16]. Others have provided similar descriptions with somewhat different names for the stages. Steve McIntosh lists the stages from start to end as: Archaic, Tribal, Warrior, Traditional, Modernist, Postmodern, and Integral[17]. These terms are the ones used in this book.

Each cultural stage develops in response to the problems created by the previous stages and produces increased capabilities. As it turns out, the cultural process is a self-organizing dynamic system of values, a dissipative structure. The implication is that change occurs due to instabilities of the structure and forms more complex structures. As the structures become more complex, the networks of communication and feedback loops also grow in complexity. It should be noted that the stages may sound as distinct steps while in reality they are a continuum. This can be likened to the color spectrum; it is a continuum of changing color with names given at points along the way in order to facilitate discussion. Each cultural stage also can have its healthy and its unhealthy aspects depending on the path an individual cultural group takes. Most stages coexist in the world today.

A brief explanation of each cultural stage follows:

> The first stage is archaic. This stage was that of the first true humans and does not exist in the world today except in infants.
>
> The next stage is tribal. This stage has fear as a significant focus, and people see the world as mysterious, threatening and under the control of spirits. There are currently only a small number of tribal people.
>
> The warrior stage is characterized by oppressive tribal control, fighting to gain control, and egocentric attitudes.
>
> The traditional stage is characterized by belief in an evil world, self-sacrifice for the good of the community, seeing everything as either black or white, and having a loyalty to the rules.
>
> The modernist stage is characterized by striving to attain wealth and status, is science and technology based. Competition and individual autonomy play a larger role.
>
> The postmodern stage is characterized by consensus decision making, sensitivity to

environment and equal rights. It is self-improvement oriented.

The integral stage is characterized by taking personal responsibility for problems. It's a balance between the scientific and religious view, an integrative view of people's opinions and world situation and is a global view.

The name "holistic" is used for what will come after the integral stage.

Studies show that for the world today, there are no archaic people and less than five percent in the tribal stage[17]. An example is the tribal regions of Afghanistan. Approximately twenty percent of the world population is in the warrior stage with about half in the traditional stage. About fifteen percent are modernist and about six percent postmodern. The integral stage is just emerging. The above distribution changes significantly when we look at the developed countries.

Paul H. Ray and Sherry Ruth Anderson have studied the postmodern population in America for more than two decades, surveying more than one hundred thousand people. They have named the postmodern culture the "**Cultural Creatives**"[18]. While about six percent of the world population is cultural creatives, in the US, just over a third of the population is now postmodern/cultural creatives, and is rapidly growing[19]. The first real impact of the new culture of "cultural creatives" occurred in America in the nineteen sixties. The integral stage is just now emerging in America, only about 50 years after the emergence of the cultural creatives. This is a very short time span compared to previous stages.

It is important to realize the effect on people and society of our collective world views and values. It's through a people's world view and values that life is experienced. People with different values have different experiences when exposed to the same fearful situations. What makes one person angry can please another. What one person is fearful about may make another feel comfortable. At the higher cultural stages, there is a greater complexity and greater ability to tolerate aspects of life that were troublesome at lower stages. At each new stage, persistent problems of the past stage are solvable.

Quantum Physics

Quantum physics actually started in the first part of the twentieth century, but was slow to advance. It presented a reality so different from the existing **paradigms** that it was strongly resisted. Even Einstein called quantum physics "spooky" and could not agree with some of the concepts. Today quantum physics' full significance is still pushing against a strong headwind. In the last several decades, it has made impressive strides and many experiments around the world have confirmed more and more of its predictions. In fact today quantum physics is the most tested and proven theory in all of science.

The pre-20th century idea of the atom being the indestructible smallest building block of matter had to change. Today, the atom is seen as having subatomic components that are not really particles but patterns of probability waves. These are not the probabilities of particles as much as they are the probabilities of interconnections and interactions. Thus subatomic particles are not really particles but interactions.[20]

Quantum physicist and Nobel laureate Frank Wilczek describes the ultimate source of mass as being made of energy[21]. In addition most of the mass of the universe (95%) is invisible "dark matter or dark energy". Wilczek also states that time has no meaning without the universe[22].

Through most of the nineteenth century, it was believed that space was filled with "ether", a substance assumed necessary for the effect of gravity or electromagnetic energy to move through space between objects. An experiment in 1887, known as the Michelson-Morley experiment, showed that ether did not exist. In time, ether was tossed out of science. Later, it was found that a basic assumption in that experiment was faulty. More recently, quantum physics has shown that empty space really is not empty at all. It is filled with quantum probability fields[23]. Out of these fields come energy and mass. In other words, there is "ether", but it's just not the substance it was thought of earlier. Quantum physics is now beyond doubt having produced unmatched predictive power compared to other scientific theories[24]. This leads us into cosmological concepts.

Cosmology

Cosmology, the study of the universe, has vastly expanded during the 20th century. It was discovered that the sun was part of a galaxy of 10 billion stars, then there were other galaxies, then clusters of galaxies, then super clusters. The "big bang" theory came to general acceptance in the late 1960s. The universe was determined to be about 13.8 billion years old[25].

What occurred in the early moments, just after the big bang, was determined theoretically using quantum physics. The universe is increasing in complexity, starting right after the big bang with the creation of the smallest subatomic particles, later stars then galaxies and larger and larger clusters of galaxies. The universe is a process of ever increasing complexity that still is continuing.

Integral Theory

Since Sir Isaac Newton, most of science was focused on physical phenomena. Everything could be explained by component parts. Throughout the 20th century, quantum physics was turning "parts" into quantum fields and probability waves. The need to consider patterns became increasingly evident. Subatomic particles are actually patterns of interconnection and not things in and of themselves.

Late in the 20th century, a systems approach that integrated the pattern and the substance approaches came into being. This is known today as "**Integral Theory**". A substance can be measured while patterns cannot. Substance is about quantity while patterns are about quality. It takes both to get the complete picture. There will be more on this subject later.

The Connection

At this point, this chapter may seem unconnected to the first three chapters. We have several more turns to take before things start to come together. In some ways, this book is like a mystery novel where several separate and apparently disconnected threads weave finally into a meaningful story in an unexpected way. Our next thread is in chapter 5 where we look at the process of change itself.

5

When you're finished changing, you're finished.
—Benjamin Franklin

How Change Happens

A purported Chinese proverb goes "If we don't change the direction we're going, we're likely to end up where we are headed". When it comes to understanding our times, we must understand change. Nature is never static, it is always in motion. Life is the same. Change is inevitable and constant, although its pace and character vary.

Change and Stress

In human lives, change is moving us within a range between growing and dying. We are never completely static. People do not generally like change because it brings uncertainty and sometimes unpleasant results. Not changing feels safer, but change happens anyway. If you have a choice between growing and dying, which would you prefer? Most will pick "growing", given the alternative. When we understand change this way, it is not as bad, even if still uncomfortable.

In times past, the world was thought of as unchanging. People grew up and lived their lives just as their parents and grandparents had. The pace of change was so slow that significant change was not usually noticeable during human life spans.

In today's world, change has been occurring faster and faster. Because of this, the last 20 years of change are not a good indicator of what the next 20 years of change will look like. Civilization is now doubling the rate of change every decade. The amount of change in the 20th century (remember the 20th century?) will be more than matched in the first quarter of the 21st century!

Change itself is changing. Change is the nature of nature. Too much change keeps us off guard and not in control and it is not being in control that induces feelings of stress.

The Science of Change

Nature mostly consists of things that fit the scientific description of "dissipative structures" (as discussed in chapter 4). Below is an expansion of this subject. These structures include the characteristics of self-organizing. A self-organizing structure seeks to maintain its status quo (**homeostasis**).

When some environmental influences (effects from the outside of the structure) stresses the system (by causing conditions the system cannot easily handle), its nature is to self-adjust in order to regain the homeostasis. It does this through internal communication networks and feedback loops. This is why the nature of humans and human societies is to resist change even though change is inevitable.

The nature of maintaining homeostasis is a critical quality of complex systems. If it did not occur, separate parts of the system could each wonder off in their own direction and system function could collapse. An example: a small town has a government, police department, emergency medical support, food stores, gas stations and schools among other services. They all are part of an interconnected network of communications with feedback loops. What would happen if there were no tendency for stability (homeostasis)? Gas stations could not be counted on for having gas at any given time and, when they did, it might not be the proper type of gas. You could not count on getting food from the market any time you needed it. If you called the fire department, it might not be there that day. You can see how such a town would soon start to unravel.

The same idea applies to an individual human. Suppose your bladder took a day or two off? Or your heart worked continuously for six months, then took an hour off? While driving a car, we expect our eyes, brain and limbs to work well all the time and not just most of the time. So, we can see why dissipative structures are self-adjusting and resistive to change.

We can also see that change is going to continue. This resistance to change in the face of continuous change is a built-in stress producer, but the stress has an important function.

For our discussion, we will break change into four types. The first is "**normal change**", the change that we are immersed in all the time. Examples are innumerable, including the minute hand of a clock, AM and PM, day and night, day of the week, weather and news. We do not usually notice normal change as such. It is always going on and we are used to it, so it does not stress us much. If we are outside and the weather changes from sunny to heavy rain, there may be a short lived stress, but it quickly passes.

The next type of change is "**small change**" (See Figure 5.1. a modified form of figure 4.2. and notice the bottom scale, labeled "Stress Factor"). Small change occurs when environmental conditions raise the stress factor a bit beyond the normal range. We feel the stress and this change takes a bit more effort than normal change. The change is not a normal one that we are used to, so it is noticeable. An example of this is when the sun rises in the morning, the light is changing, but this is normal for us and does not lead to stress. But if a lane of the road to work in the morning gets closed for a week for road work and leads to a slowdown, it may lead to a small increase in stress. We may then leave a few minutes earlier until the roadwork is done. That would be small change.

The third type of change is "**medium change**". We have seen in chapter 4, humans and human societies are dissipative structures and that includes self-adjusting. When the environment of a structure is in the "small change" range, the structure easily adjusts without much notice. But when stress rises a little beyond the small change level (3.0 for this particular example, as shown in Figure 5.1) we cannot easily adjust and reach a bifurcation point. The value 3.0 is called a critical point. If we reach a stress factor of, for example, 3.2, we are in the "medium change" mode.

When in the medium change mode, we are trying to adjust in predictable and ordinary ways, and not immediately succeeding. If the stress remains at that level, we are still seeking the old homeostasis and not finding it. We are generally dissatisfied with things, but not able to fix them. Sound familiar? If the stress factor decreases to below 3.0, things settle down and we go back into the old status quo. If we stay at the stress level of 3.2 for a period of time, we will self-organize into a more complex system that will be able to handle that particular level of stress. The critical point for that state of the system then changes from 3.0 to 3.2 (for this example) and the system is then able to easily handle a higher stress level than before.

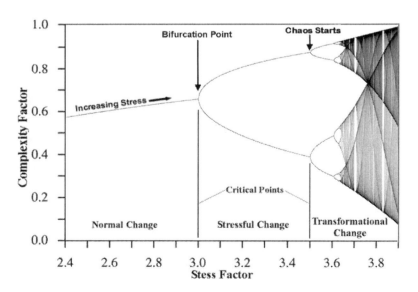

Figure 5.1. Dissipative Structures and Stress

If the stress point rises to a value of the second critical point (3.5 in figure 5.1) before the system has reorganized to a higher complexity, we enter the fourth type of change we will call "**transformational change**". This second critical point I label the "**break point**" because it is here that orderly change breaks down into chaotic change.

Transformational change is chaotic. It is unpredictable. However, as chaos theory shows, it does move within predictable regions and has some sense of order. This is a system either beginning to come apart or transforming itself into a new higher order, where it will be more complex with increased networking, communications and feedback loops that are more able to adjust to its environment.

Change produces the necessary ingredients for the banquet of life. It leads to cultural and sociological progress, and humankind can certainly use some now. Just look around at the world's state of affairs. There is not only room for improvement, a modest improvement won't really do. A big improvement is needed. A significant improvement takes a large amount of stress; it is the large stress that motivates large change, a change that will be transformational.

For example, imagine it is an exceptionally nice spring day and a person is seated on a park bench reading an exciting book. That person would not want to get up and leave at all. The minor stress of a barking dog or a couple of persistent flies may irritate a bit, but will have no chance of making the person get off the bench and leave. That would be analogous to the stress of small change. However, a life threatening event, such as a fast moving fire, would get the person up and running quickly. That would be analogous to the stress of transformational change.

How does change affect things? It does so by pushing against the status quo enough to get to the point where the system adjusts to a more complex form. Change can be beneficial even when it is resisted. We love our ruts and it is change that bumps us out, so it is uncomfortable but necessary. Change is how we move beyond our ruts. Without change, humans would still be small scattered groups of hunter-gatherers. Actually, without change, there would be no humans or life for that matter. Even the smallest single cell bacteria are dissipative structures[1].

The universe and life are dynamic not static. More complex processes and systems, however, reach periods of stability (which are not changing in some ways.) Even in stability, change is always going on at lower levels. It is constant. In larger complex systems, change on a large scale happens in stages. It goes in spurts; change, consolidate, change, consolidate, a repeating cycle like sleep, awake, sleep, awake. A tree grows in spring and summer then rests in fall and winter. There is a season for everything.

Transitions

As we've seen, change is inevitable. When looked at with this perspective, we can ask "what is dying and what is growing?" In nature, it is always one or the other. Often, when change is occurring, both are going on at the same time. When a tree is dying, it is preparing the way for new growth by becoming fertilizer and opening up space. The dead tree provides a rich habitat for new life.

The process of change is called "**transition**". Normal change occurs without significant transitions; small transitions with small change; medium transitions with medium change; and large transitions with transformational change.

When a culture is in transition, one aspect is dying while something new is birthing. Both processes bring discomfort and unpleasantness. In dying, we feel a sense of loss, we grieve. In birthing, there is pain and struggle. So it is also when a culture is in transition.

The process of transition consists of three-phases[2]. People and cultures go through each phase while in a transition. The first phase is letting go of the old ways. This phase can feel like a dying process and generally comes with a feeling of loss and sadness. The second phase is the in-between time, a time of uncertainty and stress. The third phase is the new beginning. Transition is a process of starting with the ending, and ending with the starting.

It is important during a transition to recognize the psychological need to let go of the old and to embrace the new. It is normal to resist change and feel a sense of loss when transition occurs, such as with the season of autumn, where we may not want to let summer go yet. We must deal with the ending before we can properly go on with the new. People often fail to see an ending as part of a process that leads to a beginning. Thus we may not deal with the ending and drag it into the in-between time and the new beginning, where it can limit our creativity.

We also can expect a psychological period of feeling unsettled after letting go of the old before the new birth bursts forth with its fresh creative energy. This time is like the winter season. Although this may be a time you would rather avoid, it is an important part of the process and should not be avoided. It is a time of regeneration and preparation. Spring needs the winter to prepare its way, before it ushers in a time of creativity.

This stress of transition is the driving force behind living and experiencing. It is the juice of life, the spice, the creative force. The only problem is when we fear. It is the fear of change that makes life painful.

This fear is a basic nemesis of humanity. The more we learn to live life without our habitual fear reactions, the less pain and more excitement we experience.

Being prepared for change makes change easier. It is less painful when one recognizes that change is progress and is open to it. Understanding why change happens and the process of change helps remove some of the fear and discomfort. If it has been a long hard winter and you understand the change to spring, you can look forward to the change. If it is fall and you are not looking forward to winter, you can at least accept it knowing that spring will follow.

It is change that makes life and the universe work. Charles Darwin said, "It is not the strongest of the species that survives, nor the most intelligent, but the one most responsive to change." Change is as basic to existence as you can get. It is ironic that all life is comprised of self-adaptive/self-organizing systems that seek to remain the same and resist change. Yet even here change is a basic part of life. This paradox is necessary. Without a self-adaptive/self-organizing system, there would be only chaos on a normal basis, and no complex systems such as humans could exist. Yet without a method to change and adapt, life would not occur either. There must be both, existing in balance, that makes life work. The dissipative structure with its bifurcation points is the method by which both natures come together and function.

So it is natural for us to resist change, yet change is necessary. It is also natural and necessary for environmental stress to drive us to change. This process is called evolution.

6

Chaos is the law of nature; Order is the dream of man. —Henry Adams

The Need for Today's Chaos

Most people would probably agree that today's times are chaotic. Most people would also agree that they would rather it not be nearly as chaotic as it is. In this chapter, we will discuss why we need this chaos now.

Why Chaos?

Change is one of the most stress producing things we face. In today's world we see much more of it on a regular basis than has ever been experienced before. We not only have more change, but it seems to be happening at an accelerating rate. It is not just our imagination. Change is accelerating faster and faster, and it will continue to accelerate at least for a bit more, as we shall see.

Chaos is an orderly and precise process of transformation. Its results are unpredictable to us because they can be influenced by every small input, and we don't know enough to understand how these small changes affect the results. So to us, it seems unpredictable. Chaos is transformational because a system, undergoing a chaotic period, will (if it survives) become more complex and have a greater network of inner communication that allows the system to function better. In other words, the system is transformed into a newer version with greater complexity by the chaotic process. Thus, chaos is a process of growth.

To illustrate this, the caterpillar goes through a process called metamorphosis, which is Greek for transformation. A caterpillar's function is to eat (kind of like a teenager). It can eat many times its body weight in a day and will grow to 100 times its original size in just several weeks. Soon the caterpillar gets bloated and hangs from a leaf or branch to sleep. While sleeping, the outside of its body becomes a hard surface called a chrysalis. Within the chrysalis the caterpillar's old body begins to die. Its body turns into a kind of goo. Its own immune system turns against it as it enters a period of chaos. A group of cells called 'imaginal cells" begins to multiply. These cells eventually turn into a butterfly. The new form of the creature has abilities that the caterpillar did not have, such as flight. At the same time, some of the old falls away. In this example, the caterpillar grew until a level of stress was reached that pushed it into a chaotic condition where it self-adjusted to a higher order of complexity with new capabilities. Without that chaos, there would be no butterfly.

While the description of the caterpillar metamorphosis is a more extreme example, it is useful for illustrative purposes. A simple example is the process of making a major street modification in a city to improve traffic flow. While the results are beneficial, the process looks chaotic if one does not know what is going on. Things get messy and make traffic worse before they are done, but a higher capability results in the end.

One observation we can draw about transformation is that it is preceded by a period of increasing stress leading to a bifurcation point. This is followed by a period of chaos, during which reorganization occurs.

Many people are feeling our times have been producing ever-increasing levels of stress. They feel we are now in a time of chaos. Most people can probably relate to both of these. The implication is we are now in a period of transformation, a cultural transition to a higher order.

One can think, "this is bad", or one could think, "What a great opportunity, how exciting". Both views are justified. The street is all torn up and traffic is a mess. But a traffic problem is being fixed and traffic flow will be better than it was.

The Function of Chaos

Nature has included the systems process of dissipative structures in order to make more complex systems possible. Nature also has the process of chaos to make these same homeostasis seeking structures evolve into higher forms. The juice that drives chaos is stress. It is therefore a natural and necessary part of nature. This means not only some stress, but periodically, a larger amount is needed to push beyond medium change into transformative change. Because we naturally seek homeostasis, we, as individual humans and as a culture, resist stress and are unhappy with chaos. Even as we dislike stress and chaos, it is a necessary function of nature.

Understanding this aspect of our nature does not make it feel better in chaotic times necessarily, but it can give us a sense of reason about it. It's like the "pain in the neck" street improvement project – we know it's good in the long run but don't like the stress in the moment. At least it makes some sense; it has a reason, a purpose and a promise.

Understanding the complex relationships that drive human nature gives reason and purpose where there would otherwise be worry and angst. Yes, there is a light at the end of the street project, a green light. The reason for all this detail will emerge as we go on.

While chaos is needed for transformative change, it is not for normal, small or medium change. Smaller change, like the street project, is good enough for nature most of the time. So, why are transformative change and chaos needed?

Chaos is not being able to predict or plan the change when it is underway. If we could deliberately plan and organize the change, we would not end up where we need to be. The difference is critical in nature. Humans will normally take the easiest path when given a choice. The predictable path is usually the least objectionable one. Sometimes it may be the hardest path that produces the best overall result. We, of our own volition, normally would not end up where chaos takes us. The chaotic path often leads to some unpleasant conditions before it reaches the final result such as with the caterpillar and butterfly. A business that has become inefficient over time and is facing economic failure must take some unpleasant and even harsh actions to move back to efficient and profitable conditions. Often, a new CEO is brought in to start without all the old history in the way and to make major changes that most employees would find chaotic. But, the business would not survive without the changes.

When the stress level rises beyond the break point, the message is that the old way of doing things can't handle the current situation. It requires transformation and chaos to move us out of the ruts of old thinking and into new thinking territory. That's where chaos comes in. Chaos forces a death of the old way of doing, and, if successful, leads to a new birth in ideas and a creative spurt that provides increased complexity and abilities to handle the problems created by the old rut.

Nature abhors a vacuum, and a static environment is like a vacuum. Nature is motion, and it will produce change. A glass of water sitting still and not moving is full of motion. If looked at through a powerful microscope, the molecules are not sitting still, they are moving about all the time. It's called "Brownian motion". Nature is like a restless child, it's going to involve motion.

The Cycle of Life

Chaos is part of the process of evolution. The stages of the evolutionary process are: homeostasis, increasing environmental stress, bifurcation, chaos and transformation. The transformation process is itself a process of transition (dying, neutral zone and rebirth)[1].

Homeostasis consists of growth, decay and normal change. Growth consists of a creative growth spurt, settling into a groove, and then eventually slowing to a stop. Decay is a process which goes from stagnation, after growth stops, to where pathological characteristics gain prominence, followed by the problems overtaking previous successes. This is where the environmental stress starts to build leading to the next bifurcation point. (Figure 6.1 shows the cycle of life process and Figure 6.2 shows the individual processes.)

Examples of growth, homeostasis and decay are the rise and fall of the Greek and the Roman empires; in democracies, when one political party dominates for a long period, then becomes corrupt and is replaced by another; in economics, a bull or bear market where times of growth are punctuated by recessions.

The cycle of life and nature occurs from bacteria to humans, from planets to stars and galaxies, from our world climate to continents.

Our Need for Chaos

Chaos represents an energetic state of potential. It is where things go to withdraw from a fixed state of affairs while adjusting and making room for the new. Chaos is a creative stage upon which the next act is to be performed. It is like an intermission between acts, starting with the curtain coming down on one act and ending with the curtain going back up. It is a natural part of the theater of life. Without chaos, life is a one act play.

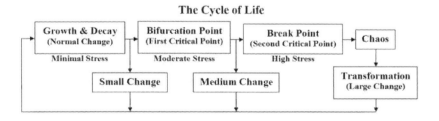

Figure 6.1. The Cycle of Life

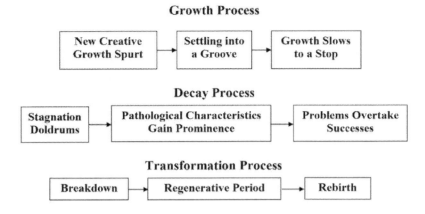

Figure 6.2. Cycle of Life Processes

Dissipative structures require chaos to move things along. Chaos makes the difference between stagnation and decay, and rebirth, revitalization and creative surges. The creative process includes times of creative spurts and times of quiet or fallow periods when we are in between. The fallow times are as important as the creative spurts. They are the times of recharging and incubation.

Change is good, not only because it bumps us out of our ruts and is the creative juice of life, moving and energizing evolution, but also because it keeps life interesting. Much of the best work creative people do is during or right after transitions.

Life is full of transitions. These include starting school, starting college, graduating, getting a job, losing a job, moving to another house, city or country, getting married or divorced, having children, losing family, experiencing bad health and dying. Life can be described as periods of homeostasis interspaced with times of transitions.

Think about books of fiction, motion pictures or television drama programs. Most of them are about life transitions. Few are about a time of homeostasis (yawn).

Our culture has interesting views about transitions. It views them from the mechanistic model, in which life is viewed as mechanical in nature. When a machine is running well (homeostasis), it is good. But if it should have a transition (fail to operate properly), it is bad and needs to be fixed. It is said to be broken.

From the first bacteria to human beings, life has been a story of transformations. Few think that transformations have stopped, that humans are the most perfect thing possible. Nature and evolution are not done. Life is an unfinished story. The transformations will continue.

So what determines when the next transformation will occur? That's a hard question, but the evidence is that it has arrived. The current chaos is the end (the first phase of a transition). The new begins down the road a bit. We have the middle phase (the neutral ground) of the transformative process to pass through. But before the neutral ground, we must first finish the ending.

We are living at rare time in history, being witness to a transformation of the human race (not its first). While it is unsettling when it's happening, it helps when we have some idea of what it's about. We have passed the bifurcation point and the stress level kept going rapidly up. We have passed the break point and moved into the chaos phase called transformation and the stress level continues to climb. The transformation is and will continue to occur.

When in time did these points happen? It does not really matter what the particular date was. We now seem to be in the chaos stage and thus beyond the point of predicting any details, all one can say is "hold on". Passing the break point is kind of like when a rollercoaster has reached the top of the first big climb and is just starting to inch forward under the tug of gravity. We don't know exactly what to expect, but we know it will be a wild ride and here it comes.

Where this coaster ride will take us, no one can say. But there are some interesting factors to help give us some clues. Remember, in chaos there is a recognizable pattern and identifiable "strange attractors" around which this pattern is organized. This will give us some important clues as to the journey at hand. But before we can examine that further, there are several more pieces of the puzzle to put in place.

The next several chapters do that. In Part III we will consider an integrated and systematic way of examining some of these "pieces".

Part III

An Integral View

7

Our brains interpret the input from our sensory organs by making a model of the world. When such a model is successful at explaining events, we tend to attribute to it the quality of reality. –Stephen Hawking & Leonard Mlodinow

Integrated Systems Thinking

Integrated systems thinking is a stuffy way of talking about the big picture of how things work. Why be concerned about the big picture? While it is important to study things in detail, like trees, somethings will be missed without looking at the larger context that the details are a part of, like the forest. The nature of a trees bark, roots and branches come from the details. But overall forest patterns, densities, habitat and climate affects come from studies of forests. It takes both details and the big picture to fully understand the story of trees. This book is not about life's details, it is about the big picture of why the world is seeing the difficulties it now is, and where it's taking us; the big picture of now.

One big picture system is called Integral Theory, developed over the last several of decades. Its chief architect is Ken Wilber although many others have contributed, and its roots go much further back in time[1,2]. While still too new to be widely accepted in scientific thinking, integral theory is making many inroads in various disciplines, ranging from medicine, sociology, psychology, archeology, economics, education, business and politics among others. Integral theory is an attempt to combine cross-cultural studies of the world's cultures, including philosophies and the great traditions, as well as many branches of science, to produce one general theory that is all-inclusive and incorporating the best of all, Western as well as non-Western. Such an ambitious undertaking has produced a relatively simple yet elegant way of bringing better understanding to various diverse subjects, traditions, cultures and scientific disciplines.

The theory produces a view of things that facilitates understanding how different ideas, cultures and philosophies fit into a common frame of reference and how they relate to each other. This picture is referred to as a "map". So integral theory has produced a map of how things are related to each other and how they fit together in the big picture. It consists of five basic factors. By bringing various differing views into an integrated whole that shows how opposing ideas are each part of one picture, integral theory demonstrates that no one has the one right view by themselves.

That helps take some of the steam out of us-versus-them thinking, and we certainly could use less of that steam in our world today. Not only does integral theory help explain opposing views in science or other life situations, it helps communication between various differing types of endeavors, such as business with ecology, education or art. The use of the same theory becomes the communication connection that otherwise would not be there.

Integral theory is about wholeness and completeness. It is about seeing things as part of the greater picture rather than looking closely at pieces as separate things all by themselves. The difference is that when seen as the whole, **emergent** qualities are identifiable, when seen as separate things, they are not. Emergent qualities are those that arise when parts join into a new whole. An example is water. It is made up of atoms of hydrogen and oxygen, elements which, when taken by themselves, have certain qualities, but not the special qualities of water (they are not wet, for instance). When the atoms are combined to make water, the special qualities emerge. Another example is carbon atoms; each shares certain qualities. But the unique qualities of a diamond do not show up (or emerge) until the atoms are joined into a crystal form (a new whole) that we call a diamond. It is the emergent qualities that differentiate a diamond from a lump of coal. You can see a world of difference between a store that sells diamonds and one that sells coal. Integral theory is like the diamond: It has emergent qualities as well as many facets.

Some of the science going into integral theory was not available just several decades ago. It is now possible to obtain university degrees in Integral Studies. Still new enough to be in the shakeout and maturing phase, integral theory is being used to provide a better understanding of life, science and cultures in many areas. The five main factors of integral theory are named Quadrants, Levels (or Stages), Lines, States and Types. These factors are maps of characteristics and interrelationships that help explain how things function and interact. This chapter gives a brief introduction to several of integral theory's major factors which will be referred to later.

Holons

Although **holons** are not one of the five factors, they are an important idea that helps explain the elements. So we first look at what holons are. The basic idea of a holon is that it is simply an individual whole thing that is also a part of something else. Each holon is a self-contained whole when considered by itself and is a part of something bigger. The term "holon" was coined by Arthur Koestler in his book The *Ghost in the Machine* in 1967[3]. Koestler had noticed in his research that most things in nature had a similar characteristic which he named "holon". A molecule is a holon. No matter how simple or how complex the molecule is, it has its own unique characteristics depending on its particular arrangement of atoms. A molecule with two hydrogen atoms and one oxygen atom is called "water" and has distinctly different chemical characteristics from a molecule with two hydrogen and two oxygen atoms (oxygen peroxide, not something you would want to drink). An atom is also a holon, being made up of electrons, protons and neutrons, which are smaller holons. Atoms are part of a natural progression of holons starting with the smallest identifiable matter and moving to molecules, cells and organisms and larger. Koestler called such an arrangement a **"holarchy"**.

The term "holarchy" was introduced to differentiate natural orders from social or cultural hierarchies, which have come to have a very negative meaning to many. The latter involves human value judgments as well as equality and fairness issues while holarchies are the structural order of nature and are without value judgments. There would be no meaning in discriminating against an atom in preference to a proton or implying an atom is superior to a proton. They are both equal and individual parts of the whole and each has its unique purpose.

In a holarchy, each stage of holons depends on the holons in the stage before it and is necessary for the holons in the stages after it. For example, a molecule depends on atoms for its existence, and without groups of molecules, the cell would not exist. If you break up the molecules, the cell ceases to exist while the atoms continue. This is one of the characteristics of holons, that at each stage in a holarchy, the holon depends on the holons in the stage before it, and supports the holons in the stage after it. The holarchy has more holons at the earlier level supporting fewer as you go toward the more complex stages. In other words, it may take a large number of atoms to make a particular molecule, and many molecules to make up one cell, and so on along the holarchy. It takes all the earlier holons working together to produce and support the later holons in a holarchy.

Unlike some human aspects of social or cultural hierarchies, holarchies are symbiotic in nature. They come about in nature because holons support each other's needs and thus benefit in holarchical relationship. Holarchies are self-organizing dissipative structures with communication and feedback loops between stages of holons in the holarchy.

A human being is also a holon. You may not be aware of it, but at subtle stages the subatomic particles in your body are in communication with the rest of your body all the time (It's probably a good thing we are not aware of this. What do you say to a proton?).

Integral theory has adopted the holon idea and with more recent information, expanded upon it. All holons share several characteristics. One is to maintain its own system (self-organizing). Another characteristic is to support maintaining the holon community that it is a part of. Another, holons at the same stage jointly make up the next stage holon. Holons also have the characteristic of moving back or forward between stages in a holarchy through the processes of bifurcation and reorganization. Each holarchy can be described as an order of increasing wholeness. All natural growth processes follows this pattern. Each holon has a complete inner communication system with feedback loops that are not present as a system in the individual parts and therefore is more than the sum of its parts. Thus a holon transcends its parts while also including them. Holons contains emergent qualities, qualities that were not expressed by their components.

The Four Quadrants

The first factor of integral theory we will examine is called the four **quadrants**. When we look at our experience with life, we can divide it into interior or exterior. Interior is what we feel, experience, think, know, and our emotions. These are all things within us. Exterior is our environment, other things, or the objects of our world. We can also divide life into individual and others. This is like singular and plural. For convenience, the word "collective" is also used for plural. Collective means more than "I", such as "we" and "them". The quadrants include everything. It is either interior or exterior, and either singular or plural. This gives us four quadrants, as shown in figure 7.1. We can label the quadrants as follows. Looking at the upper left quadrant, it is interior and singular, so we can refer to it as the "I" quadrant. The upper right quadrant is exterior and singular, so we can refer to it as "It". The lower right is exterior and plural or collective, so we can refer to it as "Its". The lower left is collective and interior, so we can refer to it as "We", shown in figure 7.2.

The quadrants can be used to categorize most anything. Let's add some meaning with examples. If we talk about a clock, it would be in the upper right "It" quadrant because it is singular and external to us. If we talk about a dream, it would belong in the upper left "I" because it is internal and singular. If we talk about our family, it would be in the lower

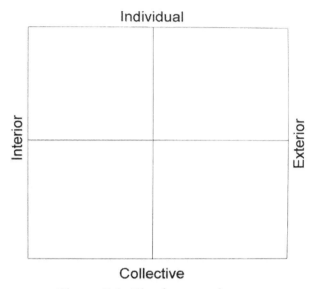

Figure 7.1. The four quadrants.

left "We" because it is collective and internal (our family is internal to us, as opposed to the rest of the collective). If we talk about clocks, it is lower right "Its" because it is external to us and collective (plural). Using an example with music, if we talk about the song "People", it would be in the lower right because it is collective and exterior. The song "For He's a Jolly Good Fellow" would go into the upper right because it is exterior and singular. The song "My Way" would go into the upper left because it is interior and singular. The song "We Are the World" would go in the lower left because it is about "we" and internal (in the sense of our world, not other worlds).

At this point, the four quadrants may not seem to be of any great significance. The value comes from a fuller understanding of how these can be used. If you are familiar with Myers-Briggs personality assessment profiles or Carl Jung's psychological types you have a general idea of some ways this approach can be useful. Integral theory is different however, and goes much further and establishes new territory.

Figure 7.2. The four quadrants with labels.

The four quadrants can also be thought of as "Intentional" for the upper left, "Behavioral" for the upper right, "Cultural" for the lower left, and "Social" for the lower right. The quadrants also represent holarchies which show up as lines.

Many different maps of relationships or activities have been developed through the years. Most are focused on the aspects of one of the quadrants. By using this approach, we can see from each quadrant's point of view and integrated things into a whole picture approach. An important feature of integral theory is its ability to take different (and often opposing) views about a subject and show how they are each valid, just in different quadrants. It is this understanding that can bring some common ground to a hotly disputed topic. A liberal and a conservative can be engaged in a lively debate, each seeing the subject from a different quadrant and each feeling very sure about their own view. But when seen from the integral view, the reasons for the differences become more apparent. So does the knowledge that both have valid points of view and neither is exclusively correct.

Lines

A "**line**" in integral theory represents a holarchy (natural progression of holons) with its start at the middle of the four quadrants and moving toward a corner. In figure 7.3, the example of a matter holarchy is shown, with the earliest stage at nearest the center of the quadrants, and moving outward.

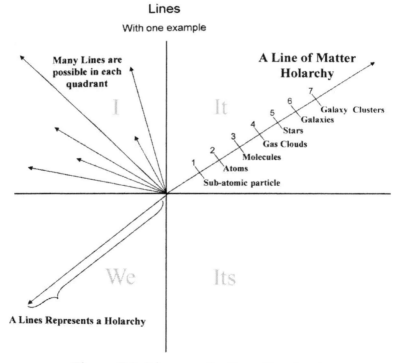

Figure 7.3. Lines on the Four Quadrants

The movement outward from the center of the quadrants represents growth or development into more complex forms of holons. Many lines are possible in each quadrant as all holarchies will fit into one of the quadrants. A given subject can have a line in each quadrant, each representing that quadrant's aspect of the subject. This is one of the powerful aspects of integral theory because it brings into view all aspects, or quadrants of a subject.

Levels or Stages

The words "Levels" and "Stages" are used interchangeably in integral theory. A **level** or **stage** is a common grouping of holons. For example, the line for the holarchy of matter in the upper right quadrant starts at subatomic particles at the lowest level (the smallest known unit of matter) then progresses along the holarchy one level at a time from atom to molecule and so on. In this holarchy, all Stars are at the same level, or stage. In figure 7.3, the holarchy of matter has seven stages. The word **depth** is used to differentiate which level of a holarchy a holon is in. Depth is measured as increasing as it moves away from the center of the quadrant. A molecule is at greater depth than an atom, but less depth than a star.

It is helpful to remember that this is a map of characteristics. The specific list of seven stages in this example does not mean to imply that this particular list is the right list and others are not. There can be many ways to define aspects of things and produce a different holarchy, with different labels and more or fewer stages. This is like a road map where there is more than one way to plot a course from point A to point B. Each way may be a valid course, but the difference provides a tool for discussing your planned travel and weighing the advantages of each course for its particular benefit. In some lines, like in the stages of matter example, the holons levels are easily defined. In other lines, like with a social systems line, they may not be so clear, and vary by what definitions are used to distinguish one from the other.

States and Types
States and **Types** are the other main elements of the theory. As we do not refer to them in our discussions, for simplicity, they will not be defined here.

Applications
In integral theory, the acronym **AQAL** is frequently used. The acronym stands for All Quadrants, All Levels. It includes all lines, types and stages that apply also. It simply means having considered and integrated all aspects of the whole picture on a subject using the full integral theory spectrum of quadrants, lines, levels, stages and types. By doing so, the relationships between one thing and others become more clear. Any limitations in a view also become clearer. For example, if an idea on a subject turns out to include just one or two quadrants, it shows up when the AQAL approach is used, this indicates that the view is part of the full picture, but does not include everything.

One example of how the AQAL approach can be used is in Management Theory. There have been several basic models for management theory in use. One is called "Theory" Y and focuses on psychological understanding in management. Another is called "Theory X" and stresses individual behavior. Another is called "Cultural Management" and stresses organizational culture. There is also one called "Systems Management" that focuses on social systems and their environment. Each system has a group that champions it and has done research supporting its approach. Proponents of each of the four theories have case histories that demonstrate why their system works and others are not as good.

When these systems are examined with the AQAL approach, each system ends up in one of the four quadrants. Theory Y fits in the upper left (I) quadrant, Theory X in the upper right (It), Cultural Management in the lower left (We) and Systems Management in the lower right (Its). This means that each system will appeal to a CEO whose personality is in the same quadrant. A company that is more aligned with a given quadrant may do better with the management system that is also in that same quadrant. Normally, within a company there will be people in each of the quadrants.

A newer management system is based on integrating the four quadrants – in other words, on integral theory – is called Holacracy™[4]. By using a balanced AQAL system, the opportunity to provide a management system that works for everyone is introduced.

Another application is in health care. Conventional Western medicine deals almost exclusively with the upper right quadrant. Its focus is about surgery, drugs, medication and behavioral modification, all upper right. Those things are definitely important, but there is a lot more to health care. Many alternative health care options deal with things like emotions, attitudes, imagery and visualization. These aspects are upper left quadrant in character. Another part of the picture is how cultural views affect health care. Things like group values, meaning of illness and judgments, as well as group support are lower left quadrant. The social systems around health care such as insurance and healthcare policies are lower right quadrant.

When creating programs or policies that affect people's healthcare, leaving out one or more quadrant can lead to gaps or difficulties in the system. That's where using the AQAL technique can aid in assuring a rounded program where everything is integrated into a unified program rather than different programs for different quadrants.

Integral Theory's Significance

When all is said and done, integral theory is a good tool that helps bring balance into various outlooks. We all have our quadrant of personality and it affects our world view and reality as well as how we experience life and react to events. It helps us remember that there are other viewpoints/realities that are equally valid. It broadens our appreciation of others' views and when applied on a regular basis, helps us to include the others in our thinking. This ultimately improves communication, reduces disagreements and provides for overall better understanding of the variations in life. In other words, using integral theory provides for better feedback loops within the human dissipative structure holarchy, and improves holon to holon communications – exactly what is needed for the next emergent step.

To be sure, there is a lot more to integral theory than can be put in this one chapter, or even a whole book. This is not a full account of what it is, just an introduction into the basics so we can use several of its ideas in the chapters that follow.

8

I was a young man with uninformed ideas. I threw out queries, suggestions, wondering all the time over everything; and to my astonishment the ideas took like wildfire. People made a religion of them. – Charles Darwin

Evolution, the Big Picture

What on Earth does evolution have to do with the state of affairs today? Evolution is the name given to nature's process of change. Like most things in nature, it is a process that follows some general behavioral rules. Understanding these rules can be helpful in appreciating what is happening now, why it's the way it is and where it is taking us. Like any field of science, understanding leads to being able to use nature's processes to our advantage. By a better appreciation of evolution, we move from being like passengers on a bus, with no idea where the bus is going, to being the driver of the bus. We can now make intelligent choices about our route and avoid some of the bigger obstacles. If we know there is a bridge out on one highway, we can take a different course rather than ending up at a dead end. Most of humanity would rather not end up at an evolutionary dead end, like the Neanderthals.

This chapter is the story about the evolution of the universe and everything in it. The reasons we are looking at universal evolution is that it includes us, is still evolving, and it is here, curiously enough, that the pieces of this story start to come together.

Not Your Darwin's Evolution

First of all, we are not talking about the evolution of Charles Darwin or survival of the fittest. Darwin's paradigm changing book *On the Origin of Species* was published in 1859. Most of today's science was unknown in his day. Many new fields of science have since developed and vast quantities of new information have been acquired. It is not a criticism of Darwin to point out that some of his ideas have been overturned. He did a great job with the information he had at the time.

In this chapter, we are talking about evolution from the view of today's leading edge in science and integral theory, a universal view. In Darwin's days it was believed small changes could account for everything. But over the years, it became clear that something more was needed to account for some of the evidence. There were discontinuities in fossil evidence that pointed to leaps of evolutionary change, change that occurred in big steps without a record of slow step by step process. Some of these discontinuities occurred at noticeable times where major environmental stress was occurring. Examples are ice ages, massive volcanic eruption phases, and asteroid impacts with earth. In other words, times of great disruption and chaos. Darwin's view is still useful for some of the smaller evolutionary changes that go on in the day-to-day world while a different view is useful for the larger changes, such as the introduction of new species. That is where chaos comes in.

Two key aspects of Darwin's ideas were evolution through gradual change due to random mutation, and natural selection (often referred to as "survival of the fittest")[1, 2]. The later addition of genetics produced what is now called Neo-Darwinism. This is the view taught in many schools around the world even though it is outdated. Newer science of the last few decades has shown that these two ideas are inaccurate. One finding has been that most evolution has not occurred gradually over time due to the accumulation of genetic mutations. There have been long periods of stability followed by sudden significant changes[3, 4].

When looked at from a systems theory point of view, the genome is seen as a self-organizing system with internal networks that are able to spontaneously produce new forms. This is in stark contrast to earlier views (called genetic determinism) of genes being passive molecules that change only through mutation, and control all of our physical traits as well as emotions and behaviors.

In the mid-1960s, it was thought that a human being contained over six-million genes. That number lowered to about one-hundred-thousand by 1990 when the massive human genome project was started to identify each gene. It was assumed that each gene made a single type of protein and there were that many human proteins identified. When the project was finished in 2003, only about 23,000 human genes had been identified. This was humbling as the common grape has over 30,000 genes, sour or not[5]. New research suggests that there is not one answer because the number of human genes differs from person to person[6].

The genome project showed that the assumptions about how genes worked were incorrect. Newer studies have indicated that the genes do not control human traits. The genes produce parts of proteins that are assembled into the full protein, but it is not the gene that decides as previously thought. It turns out the cell's environment controls what proteins are produced[7]. When a need arises for a protein, it is communicated to the cell and the cell produces the proper protein.

When a high level of environmental stress is encountered, the cell goes into a mode of producing mutations[8]. The ones that work better in the stress environment are replicated while the ones that don't are not. Thus, mutations and natural selection work through the cell, but only when the need arises due to environment stress. This is the process of a dissipative structure after reaching a bifurcation point and entering the chaos phase of change. This also explains why evolutionary "quantum leaps" have occurred during times of great stress like ice ages and mass extinctions. The mutations produced however are not totally random. Studies on different batches of identical bacteria separately produce similar end results. This is not surprising when we remember that in chaos, there is a degree of randomness present, but it is within a given pattern.

Darwin's idea of natural selection later came to be referred to as survival of the fittest and led to the idea of a dog-eat-dog world, where nature was competition of one against others. This us-versus-them thinking is prevalent today throughout our culture. In more recent times, a vastly greater understanding of the process of life and evolution has shown this view is only a part of the large scale history of life. From the very early forms of life up to humans, it is cooperation and **symbiosis** (the word symbiosis can have several meanings, here it is taken in the original cooperation sense) that are the story in successful living systems. Although competition exists in the sense that each living thing works to acquire its needs in a limited environment of supply, in nature it does not usually result in one system working deliberately against another to gain advantage. The competition is more a byproduct of active seeking than the deliberate strategy common currently in human culture. The most successful living organisms have been ones where working together with other organisms have produced a cooperative environment to the mutual benefit of both[9]. Complex systems cannot exist without substantial levels of internal cooperation. In fact, natural science is learning that cooperation is a fundamental evolutionary principle[10, 11]. It's a far cry from the old negative dog-eat-dog attitude even with competition for resources.

We have gone into all this talk about Darwin and his ideas because there is still a lot of misconception about old ideas that no longer have scientific support, even though some are still being taught in schools. To move past those ideas, it was important to get them out of our way. Now let's move forward into the past.

Growing a Universe

How do you grow a universe? Start off with a bang; a big one. Cosmologists talk about the evolution of the universe starting with the Big Bang. To look at that we need to go back, way back in time, so far back it was before time, at least as we know it. Looking back 13,820,000,000 years, give or take a few, we are at the moment of the Big Bang. In a manner of speaking, it was before space, time or matter. There was only, well, nothing.

At the instant of the Big Bang there was what science calls a **singularity**. According to Stephen Hawking in his book *A Brief History of Time*, even the laws of physics don't go there. It just was. There is no practical sense in trying to discuss what it means at that point or what caused it, as it is beyond current scientific understanding. There are some who theorize and speculate about the origins of the big bang, but at this point, they are more or less wild ideas, involving other universes, dimensions and ideas given names like strings, bubbles, baines, bounces and bangs (no, it's not a wild party). Fortunately, we need only to start with the Big Bang and not worry ourselves about why it happened or who was to blame.

This story of evolution really begins at an infinitesimally small moment just after the Big Bang, when space, time and the laws of physics have just come into existence, even if just barely. (At the first moment in time, there was an unimaginably large flow of energy from the point where the singularity had been that became the start of the universe.) There was no matter in the new universe and the energy was exploding outward at incredible speeds. Although there was no matter yet, there was a vast quantum field consisting of waves of potential along with an incredibly high density of energy. As the new universe expanded, the energy density decreased. A short time after the bang, the first building blocks of matter condensed out of the energy flow.

When talking about matter, one must first define what matter is. That can be a complicated task. For our consideration, we will call matter something with mass that can be identified and specifically located. An atom fits this description. But so does the atomic nucleus which is made up of protons and neutrons. There is a whole zoo of other smaller objects (electrons, quarks, gluons, partons, hadrons, mesons, baryons, bosons, axions, leptons and maybe klingons). Frank Wilczek, noble prize winner in physics, refers to these (except klingons) as different from matter in the usual sense, more like "embodied ideas" than things.[12].

The first things to condense out of the cooling universe were the zoo of embodied ideas. There is quite a bit of theory on what went on in the zoo that we won't go into here. This all occurs in less than the first second of time. At about three minutes into time, the first protons and neutrons form into atomic nuclei. At about 300,000 years after the Big Bang (**ABB**), the first stable atoms form[13]. The vast majority were hydrogen atoms, with some helium and maybe a bit of lithium. At about the same time, the cooling universe entered its dark ages. The fading light from the Big Bang receded into a faint background glow and the cosmos went dark[14].

At this point I should mention one other aspect – dark matter and energy. Science is still struggling with this concept. It appears that most of today's universe may be made up of something that can't be seen, thus the name "dark matter". According to calculations, this unseen stuff must make up over 90% of the universe to account for the motion of what we can see. Although evidence is growing in support of dark matter and dark energy, no one knows for sure if it really exists, and if it does what it actually is. In other words, to explain the motion of the visible universe, science has invented something with mass that can't be seen, has never been measured directly, and has unknown characteristics. No, this is not a Harry Potter story line. Since we can't see it and can't measure it, if it indeed exists, let's do the practical thing and ignore it. Just think of it as one huge skeleton in the closet, and we won't open the door. Although important to astrophysicists, it really does not make any difference to this story whether it's there or not. On with the story.

Slowly, individual atoms grouped together in various regions, attracted by their mutual gravitational force. In time, these clumps grew larger into clouds of mostly hydrogen gas. As they moved inward from gravitational attraction, they started to rotate. At about 300,000,000 years ABB, rotating gas clouds got so massive and dense that the internal pressure and temperature reached a critical point and a thermonuclear fusion reaction started. A star was born. The universe lit up.

At first there were a few and then in rapidly growing numbers, more stars flared into life. At about 700,000,000 years ABB, the first galaxies of stars formed. By 3,000,000,000 years ABB most known galaxies had come into existence, totaling more than 100-billion, each galaxy having billions of stars (that's well over 100,000,000,000,000,000,000,000 stars in the cosmos).

Because the early cosmos was mostly hydrogen with a little helium, any planets that formed around early stars were "gaseous" planets (like Jupiter) not "rocky" planets (like Earth). All the heavier elements (like oxygen, carbon and iron) did not yet exist.

The life cycle of a star covers a number of changes as it ages. As a star first flares into life, it starts producing energy by thermonuclear fusion (the same process as in a hydrogen bomb). This process fuses two hydrogen atoms together producing a helium atom and releasing energy. The star does not blow up because the incredible mass of the star contains the force and also stabilizes the rate of burn.

The star enters old age when it has used up its hydrogen. The star then changes and starts fusing helium into carbon. When it has used up its helium, it fuses carbon into neon, then neon into oxygen, then oxygen into silicon and finally silicon into iron. Not all stars go through this entire cycle, it depends on their size. Cosmologists have found that stars come in a variety of types and sizes and each has a unique life cycle. Smaller stars die out after their iron is used up. However with stars several times larger than our sun, a very important evolutionary step occurs when the iron is used up. The large star suddenly collapses, producing an intense reaction that results in a **supernova**. The immense explosion blows the star's outer parts (the part full of all those non-hydrogen elements) into space at tremendous speeds[15]. The explosion process also modifies some of the non-hydrogen atoms into the other elements of nature.

This process of star birth, producing heavier elements, supernova and dispersal of those elements has been going on for billions of years. Slowly, the interstellar clouds of hydrogen gas have become enriched with all the elements of matter we know and find on Earth. Most of our planet, including most of you and me, are made of elements created in stars within our Milky Way galaxy. The rest came from the Big Bang boom.

About 4.5 billion years ago (about 9.3 billion years ABB) an interstellar cloud that had condensed into a proto star ignited, producing our sun. When the sun was condensing into a tight rotating mass, it produced a large equatorial disc of matter, somewhat like the rings of Saturn, but much denser. As the sun began emitting large quantities of energy, the lighter elements in the disk were blown outward while the heavier elements stayed closer in. Lumps began to form within this disc and started collecting more of the mass in that region by gravitational attraction making planetesimals (sub-planet size objects). Those eventually accumulated into our current planets. The lighter elements in the outer reaches of the disc formed the "gaseous" planets of Jupiter, Saturn, Neptune and Uranus. The heavier elements in the inner reaches formed the "rocky" planets of Mercury, Venus, Earth and Mars (Pluto has been re-categorized as a planetesimal).

Early Earth was a place of great turmoil. Planetesimals and meteorites bombarded its surface and volcanism was plentiful. Eventually, things settled down some and as the atmosphere cooled, water started to condense, producing rain. Oceans formed. With the oceans, fiery volcanoes, dry land and electrical lightning storms, the array of star made elements formed a chemistry experiment of great variety. More complex chemical combination produced all the building blocks of life[16]. This process, from the formation of our planet to the brink of life, lasted for about six-hundred-million years, plenty of time for advanced chemical formation.

Life Begins

About four billion years ago, life appeared on Earth. The questions "what is life?" and "how did it begin?" are frequently asked. These are both difficult to answer but what follows may suffice. The definition of "life" is not universally agreed upon. For our purpose I will define life as a system that possesses four characteristics: metabolism, reproduction, adaptation and having much more complexity than non-living things[17]. The latter condition is a general observation more than a defining one as it is a circular argument (it's living because it is more complex than non-living things, and non-living things are non-living because they are less complex). Metabolism is necessary to provide the energy needed for complex systems. Reproduction is needed to continue a life form and it must be adaptive to its environment so it can change as conditions warrant.

How did life start? It appears that the evolutionary process kept moving to more complex systems until one day, it met the definition of life. In other words, life was a natural result of the evolutionary process toward higher order and complexity. While there is much room for debate on this answer, it should be good enough for our purpose here.

The conditions needed for life, at least as we know it, include the presence of carbon, nitrogen, oxygen, hydrogen, sulfur and phosphorous arranged in a wide variety of chemical compounds (such as amino acids and proteins), sunlight and relatively narrow temperature ranges[18]. Once all these conditions were met, life emerged.

For about two-billion years, the only life on Earth was single cell bacteria (called **Prokaryotes**). These first bacteria were small, simple single cell creatures that lived completely independent lives. A cell is a living organism encased in a thin membrane that separates the cell from its environment. The membrane lets nutrients in and waste products out. So it is the cell's membrane that interacts with its environment. Early bacteria found food plentiful and reproduced rapidly. We have single cell bacteria today.

These early bacteria were so successful that about 3.7-billion years ago they had consumed most of the available chemical food and outpaced nature's ability to re-supply[19]. They were living beyond their sustainable level and a crisis of overpopulation, resulting in a food shortage, occurred.

The ensuing famine caused high levels of environmental stress, a bifurcation point and chaos, leading to rapid bacterial experimentation through deliberate random mutations. This chaotic period led to a major evolutionary transformation for life on Earth.

Bacteria, use to easy living on the abundant chemical stew of early Earth, now had to learn to produce their own food or die. Some bacteria learned how to produce nitrogen by chemically extracting it from compounds that contain it. They learned to remove carbon from carbon dioxide. They also learned photosynthesis, the process of making food and energy from sun light and other chemicals. This crisis also produced the first "breathers", bacteria that take in a gas (carbon dioxide) to use in their chemical processes.

As free hydrogen began to become scarce about 3.6 billion years ago, another crisis was presented to bacteria[20]. The result was a new form called "bluegreen" bacteria. This was another major evolutionary step. Bluegreen bacteria learned how to combine two forms of photosynthesis and use carbon dioxide, water and salt to make sugars for food. In doing this, they got hydrogen by removing it from H_2O (water), leaving O_2 (oxygen) as a waste product. For a long time, oxygen waste combined (oxidized) with rocks and was not overly abundant in the atmosphere. Oxygen combined with iron produced iron oxide, which settled to the bottom of ancient oceans. Today's iron mines dig this same iron oxide out to make iron. The layers of iron oxide mined are the results of bacterial action over millions of years.

Between 2.1 and 2.3 billion years ago, oxygen had built up in the Earth's atmosphere to toxic levels. A new crisis was at hand. More oxygen had advantages. It made for a thicker atmosphere and provided a protective ozone layer, but too much led to toxic air for many bacteria[21]. Once again there were high levels of environmental stress which led to chaos.

Bacteria responded to this new crisis in several ways. The most notable was to produce a new powerful way to transform energy while solving the problem of excess oxygen. Aerobic respiration (the taking in of oxygen and using it to break down molecules for food) resulted in producing carbon dioxide as a waste product. This major evolutionary step provided for an automatic stratagem to balance oxygen production with carbon dioxide production. At the same time the aerobic respiration process proved to be much more efficient than the previous process. That energized life.

Multicellular and Beyond

For the first two billion years on earth, life consisted of single cell life such as bacteria. These cells had "skins" of thin membranes that if broken, led to the death of the cell. The bigger the cell, the more fragile it's membrane. Thus, the size and complexity of bacteria was limited.

At about two billion years ago, another major evolutionary "leap" occurred[21]. Groups of different types of bacteria formed symbiotic relationships that improved their ability to survive. These communities of bacteria had developed specialization, with bacteria that produced something usable for other bacteria. Each got from the others what it needed. In some of these communities, a portion of the DNA was stored in a central location, which became a cell nucleus. These cells (called **Eukaryotes**) evolved from these communities of single cell bacteria that had developed specialization in symbiotic relationships to their community. When some of these communities developed a common cell wall, multicellular life emerged. A higher order holon with emergent qualities was born. All plants and animals are this multicellular type of life. All life on earth consists of either single cell or multicellular forms, and each multicellular life form is made up of a collection of single cell life working together[22].

These new multicellular and nucleated cells worked more efficiently, had greater complexity, were able to develop wider diversity and were more robust than single cell life. With single cell life, evolution was limited to new species of cells. With nucleated multicellular life, evolution could proceed through the way the cells were organized, which was quicker and led to much more complexity, variation and to evolution speeding up.

One way of thinking about this is to compare a medieval home with a single cell. If a group of these homes were built together, a community was formed. In such a community, people living in individual homes could become specialist in a trade, such as blacksmith or carpenter, and become more proficient at their specialty than could any individual jack-of-all trades. If this community built a defensive wall around itself with a castle in the center, it would be in some ways like a multicellular life form. In this example, evolution could occur through the way the community functioned and was arranged without the people in each home being genetically changed. The nucleated multicellular creatures were the foundation for the future of life all the way up to humans.

After about another half a billion years (1.5 billion years ago), another major evolutionary step was made. Sex came into being (yes!). Up to that time, reproduction occurred by **mitosis**, where a cell splits in two[23]. This is really like making a clone of oneself, with the opportunity for variation being small. With sexual reproduction, two creatures came together and after the first date, shared genetic information to produce offspring with a mixture of characteristics based on the gene combination[24]. The opportunity for variation through mixing of genetic combinations was vastly increased over reproduction by mitosis.

As evolution continued organs appeared, sensing of the environment increased and simple nervous systems developed to increase communications within the organism[25]. All these helped speed the pace of evolution.

Animals arrived on the scene about 600 million years ago, living in the sea. They appeared on dry land about 415 million years ago[26]. Adapting to life on land was a significant evolutionary step for animals.

Once animals developed spinal cords and skulls to protect delicate nervous systems, these systems made significant advances leading to brains.

About 225 million years ago, mammals appeared in the dinosaur dominated world[27]. With mammals, nervous systems rapidly grew in complexity and brain size rapidly expanded. About 65 million years ago, an asteroid struck earth with worldwide results (massive volcanism may have played a part also). A mass extinction occurred with dinosaurs disappearing from the scene. As life made a comeback, mammals expanded without the constant threat of large dinosaurs, resulting in their evolutionary proliferation.

About 35 million years ago, monkeys emerged, and at 20 million years ago, the orangutans, gorillas and chimpanzees emerged. These are human's immediate ancestors, so the evolutionary path to humankind had been prepared.

Humans

When we try to address the question of when did the first human appear, we run into a morass of different theories and opinions. Science cannot answer that question today with precision. Even providing a clear definition of whom or what was human is difficult. For the following description, I have taken some liberties with dates and definitions in an effort to present a relatively simple story without belaboring all the various opinions. For the purpose of the overall saga of evolution, those details are not important. With that caveat in mind, we start with our early human ancestors.

The first species to be given the name "**Homo**", meaning human, was Homo habilis (skillful human) who originated about 2 million years ago. Homo habilis was a descendant of apes and more ape-like in appearance than a modern human, still walking on all fours. It has been put into the genus Homo because it had a much larger brain than its ape cousins and used rudimentary tools[28].

At about 1.6 million years ago, Homo erectus (upright human) walked onto the scene with a still larger brain. These were the first to start the human migrations out of tropical Africa, but still not genetically modern humans.

At this point in the story of evolution we move from billions and millions of years down into thousands of years. About 400,000 years ago, Homo **sapiens** (self-aware human) began to develop, reaching the fully developed stage about 100,000 years ago. They were close to modern humans genetically.

Homo sapiens sapiens (referred to as "modern" humans and genetically identical to us – in other words, us) developed about 40,000 years ago. It was at that time that complex language, art and music entered the scene. For the sake of this account of evolution, we will take this (40,000 years ago) as the start of the true humans as we know them. Again, dates and precise definitions are still fluid in the scientific literature[29].

In evolution's time scale, the human brain evolved at an extraordinarily rapid pace into the most complex structure we know in the universe[30]. Our brains are several times larger than what is needed for biological functioning[31]. This represents a major evolutionary step. The expanded brain capacity allowed for not only the development of complex language, which includes symbolic concepts, but abstract thinking. These are qualities unique to humans and place humans at the top of the evolutionary process (at least on Earth – there's a big universe out there).

Language allowed for detailed communication between people, which made it possible for the sharing of learning and abstract ideas so each person did not have to start at zero and learn everything for themselves. Humans are the only known species that can communicate their thoughts to each another. The advanced cognitive abilities of the human mind have resulted in the arts, music, philosophy, civilizations, science and technology and zombie movies.

These unique human qualities paved the way for a new avenue for evolution to advance – culture. Vastly faster than genetic change or cellular organization, human culture has taken on the leading edge of evolutionary progress.

Human cultures are self-adjusting, self-organizing dissipative structures as discussed in chapter four. As they have progressed, they have increased in internal communications, feedback loops and complexity. Communication has progressed from word of mouth to writing, to various electronic forms. The speed of communication as well as the amount of information has vastly increased.

The evolutionary stages of human culture were outlined in chapter four. By the definitions used, there have been six stages to date with the seventh currently emerging. The evolution of the universe can be broken down into three broad categories: the evolution of matter, life (or biology), and humans (or intelligence). Each is still in progress. The era of matter lasted almost 10 billion years, the era of biology, about four billion years and the era of intelligence, about two million years (including all genus homo). See figure 8.1.

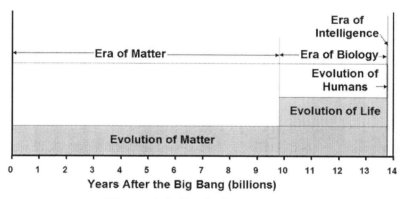

Figure 8.1. Evolutionary eras

It could be argued that the human era should be part of the biology era. But human intelligence and communication have added a higher level of evolution not existing in the pre-human world and thus deserves the separate category. This does not make us humans superior in a value judgment way.

It is because we are evolving through cultural development and not just through genetic variation of the biology era.

Evolution's Trends

If we take a systems theory view of evolution, with some added integral views, we can begin to see some common themes in what has been occurring for the past 13.8-billion years (recently bumped up from 13.7 billion years)[32]. First of all, there is a trend of ever increasing order, complexity and integration. Starting off with a burst of raw energy, the universe produced the building blocks of matter, the zoo of "embedded ideas", like quarks. Then came matter, starting with protons and neutrons, then atoms, then in turn, hydrogen clouds, stars, galaxies, heavier elements, rocky planets, the chemistry of life, single cell living bacteria, food making bacteria, breathing and photosynthesizing bacteria, multi-cellular life, plants and animals, mammals, humans, tribes, kingdoms, civilizations, globalization, and finally, fast food restaurants.

Each step presents an increasingly complex system structure with increasing order, complexity, internal communication, feedback loop networks and cooperation. In short, we have a grand holarchy of holons, a massive dissipative structure. And, there is absolutely no reason to think this evolutionary process has suddenly stopped. In fact, by its very nature, the universe must continue to evolve, or decay and die.

This is not to say that at times evolution has not taken a step back to a less complex form or that it is a steady, uniform or linear path. Evolution has been more of a meandering in a general direction. It is non-linear, but overall, it does trend in increasing order, complexity and cooperation. With an understanding of chaos theory and dissipative structures, this makes good sense. Chaos is non-linear and is unpredictable at the detail level while still following the trends of the system. Evolution does the same. While smaller steps in evolution are somewhat predictable and understandable, the big steps are chaotic in nature.

Cooperation is a fundamental aspect of evolution and is partly responsible for higher orders of complexity. A sterling example is the human body. The average adult person contains more than 70 trillion living cells. While trillions are "human cells" (containing our unique DNA) most are non-human microbes (bacteria, viruses, fungi, archaea and others)[33]. We are indeed complex societies of living cells. Each person has over ten-thousand times more living cells than our world has people. We are full of cooperative activities that make such complexity possible, without which our bodies would cease to function within minutes.

While the evolution of matter and life continue, it is the evolution of human intelligence and cultural development where the real action is today. In the next chapter, we will take a more detailed look at the trends in human evolution and where we appear to be headed.

9

The Destiny of Man is to unite, not to divide. If you keep on dividing you end up as a collection of monkeys throwing nuts at each other out of separate trees. –T.H. White

Where We Are Now

In this chapter we are going to examine where we are in our human evolution. Non-humans may skip this part. The first step is to look more closely at where we humans have been in order to uncover the trends we are experiencing today. This gives us the long view of the trends rather than our usual view based on the last decade or two, or often only the last several years.

    We will use the terminology of the chapter 4 section titled "Cultural Development Theory". The evolution of cultural development is significant because cognitive thinking developed and reached a major new level with the development of the human brain. Complex language, abstract and symbolic thinking and communication were new to the known universe. A key difference between humans and other animals is that in successive generations, humans can show significant increases in mental complexity whereas with other animals this is not as evident. Through complex and symbolic language, humans can learn by the exchange of abstract ideas and gain understanding beyond their personal experience. It is the increasing cultural complexity that represents a key differentiating characteristic of humans[1]. A new method of evolution through cultural development opened up as the result. That in turn revved up the speed of evolution. These views are the result of a massive quantity of research by numerous teams of scientists over decades in a number of disciplines[2].

Stages of Human Cultural Development

Early humans developed about 400,000 years ago. The first recognized level of human cultural development is called "archaic". It started about 100,000 years ago when Homo sapiens were fully developed, and lasted to 40,000 years ago, the advent of "modern humans". The people of the archaic development stage were hunter-gatherers occupied with staying alive and the needs of the physical body. Actions were driven by sensory inputs, more of an automatic state in support of physiological needs and survival, at least as far as we know[3]. Fire was controlled and simple tools used.

Humans formed bands, more like a herd of animals than an organized tribe. These unstructured bands increased survival and helped in procreation.

The next level was the tribal stage, starting about 40,000 years ago. The bands of archaic people had grown into small organized tribes of up to 40 members with an average life span of approximately 22.5 years[4]. Art, music, dance and ritual were developed. A high degree of community was present with strong family and tribal/clan bonds[5].

Living cooperatively and thinking of the good of the tribe were new cognitive expressions of the tribal stage people. They also began to think about what caused things to happen and that resulted in a high degree of superstition, with amulets, lucky charms and the belief in magic and spirits. Shamanism and sorcery flourished as did legends, myths, traditions, omens and spells.

Also new were "world" views. Although simple, they provided models of reality that gave room for collective tribal memories and for limited planning for the future. Another new idea was "us versus them". Those in the tribe were us, those outside were not, and were not given the same value. This led to warfare between tribes. Much effort was spent in working to keep the spirits mollified and avoiding taboos. Life was one of fear. Social interaction and structure were brought into existence by the tribal stage.

The warrior stage came next starting about 10,000 years ago. After tens of thousands of years, the tribal stage had discovered rudimentary farming or horticulture. This consisted of planting seeds using a hoe-like digging tool in limited areas.

As time went on, farming became slowly more sophisticated, producing a more predictable food supply and allowing for larger tribes, leading to the transition to the warrior stage. Expanding populations came into conflict with neighboring tribes, leading to increasing aggression.

The warrior stage pushed the tribal development into a new form. For one thing the forces of production and distribution of food changed the social structure. The tribal chief became a war lord. Individual strength and boldness replaced duty to the tribe. Competition and survival of the fittest became the rule as territorial expansion produced exploration and exploitation of others[6].

Over several thousand years, simple horticulture gave way to the agrarian revolution and the animal driven plow, leading to even larger empires and the traditionalist stage.

The transition to the traditionalist stage started about 3000 years ago in response to the harsh condition created by the warrior stage. The traditionalist stage perceived the world as basically evil and longed for order and stability. It was community oriented where the individual had to conform to the order of society.

People of this stage held reputation and honor to be extremely important and blamed whatever happened on others. It also was the stage where the population was divided into haves and have-nots, the strong and the weak.

The traditionalist's had a divine plan that established people in their place in life and provided the order and stability so desired, bringing a sense of meaning and purpose to life[7]. People willingly accepted their lot in life and conformed to authority.

The traditionalist had characteristics of fundamentalism, black and white thinking, being right or wrong and being part of the preferred group. Proper social roles, castes, classes and races were accepted without question. Although traditionalist's ways were often rigid and dogmatic, they provided for a sense of belonging and order that was deeply satisfying. Traditionalists brought writing into the world[8].

The next cultural level to develop was the modernist, coming into prominence about 350 years ago. With the rational-based thinking of science, the modernist culture broke free of the conformist and the restricted religious environments of the traditionalist. This resulted in individual rights, autonomy and independence. Individual experience replaced religious dogma.

Progress and improvement became honored. The scientific method was established and supported. Democracy, freedom of the press and speech, and equality under the law developed and blossomed.

With science came technology and the industrial revolution. That in turn led to a middle class rather than just the rich and poor. Modernists are individuals responsible for their own success and in competition with everyone else for the good life. They are materialistic and driven to succeed. They are status and fashion oriented. Ethics can easily be overlooked in the race to succeed and short term gain outweighs long term gain. Modernists want it now. They push hard to get on top and stay there and may use others in order to win the competition.

Modernists introduced modern science, industrialization, technology, democracy and global climate change into the culture.

The postmodern level burst forth in the 1960s. This stage resists competition, social status and labeling of people while embracing diversity, equal rights, cooperation and sharing of resources. Non-violent change and consensus are popular while social hierarchies are not.

Diversity is sought in cultures, race, religions, gender, philosophies and health care. The postmodernists are suspicious of membership or organizations. They feel everyone is equal and important in their own right and want love and peace for all. They are big on cooperative community, especially if it is not too well organized. Postmodernists tend to be very tolerant of others and sometimes are seen as bleeding heart types by non-postmodern.

Postmodernists brought us civil rights, feminism, ecology, the information age, global networking and sit-ins. They also brought alternative medicine and a return to whole grain bread.

The next level is the integral stage, now emerging. At the time of this writing, it has not yet blossomed into a mainstream reality. However, its presence is felt as it gathers in the wings on the stage of life. Before discussing what the integral stage looks like, it would help to examine some of the common characteristics or trends of human cultural development to date.

The established seven stages of cultural development are summarized in table 9.1. One trend is the cyclical movement back and forth between "self-oriented" and "group-oriented" patterns of behavior. Each new stage has been the opposite of the previous one. Modernists are self-oriented while postmodernists are group-oriented. Because each new stage has greater depth and complexity and contains the oscillating behavior between self and group orientation, the term "Spiral Dynamics" was created to describe the progression of human cultures over time[9].

When talking about "higher" stages of cultural development, or more advanced stages, a false sense of value, or one being better than, or superior to another may arise. This is not really a helpful way of thinking about the different stages. A given stage enfolds the earlier stages and does not exist separately. A more meaningful way to consider the stages of development is to think of them as nested spheres. The earliest stage is in the middle and each newer stage develops from, and includes the one before it. Figure 9.1 depicts this. Thus, each stage is fully dependent on all previous stages. The human body can be used as an analogy, with the skeleton at the center of the body while the skin is the outermost layer. The skin is not superior to the skeleton and both are fully required for the body to function.

Table 9.1. Stages of Human Cultural Development

Order	Stage Name	Stage Start* (years ago)
0	Early Humans	400,000
1	Archaic	100,000
2	Tribal	40,000
3	Warrior	10,000
4	Traditionalist	3000
5	Modernist	350
6	Postmodern	50
7	Integral	Emerging now

* The start is the time that a stage becomes widely recognized

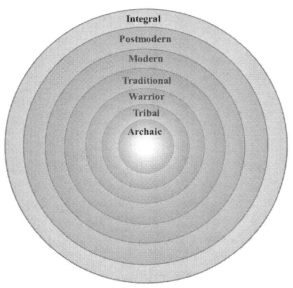

Figure 9.1. The nested spheres of cultural development, an integrated whole.

Each new stage did not just suddenly start. New stages where preceded by decades or centuries with philosophers or thinkers who laid the groundwork with ideas. Slowly over time small groups of people formed around the ideas and then eventually grew larger and began to have more of an influence. At some point a "critical mass" was reached where the influence spread rapidly and became common knowledge[10]. That event is labeled as the start of the new culture, where most of the old cultures were still active. Different individuals held world views and values in one or the other culture, or in transition between two. As an example, Ralph Waldo Emerson (1803-1882) was an early postmodernist although the postmodern culture did not make itself felt until the 1960s. Today, in the US, we have mostly traditional, modern, and postmodern cultures intermingled. There is still a handful in the tribal and warrior cultures while there is also the beginning of an integral culture. The traditional and modern cultures numbers are in decline while the postmodern and integral are increasing. The numbers reflect the overall trend toward increasing depth and complexity of human society.

One significant trend can be seen in table 9.1. The time between stages has not only been getting shorter, but the rate of change has been increasing. As we will see later, this is a trend shared by many other indicators of change.

Another trend has been noted from the history of human culture change in that a change does not occur without significant stress and a time of chaos. This fits the dissipative structure and chaos theories covered in chapter 4.

The Maturing Human Race

As the twentieth century progressed, life spans substantially lengthened and the quality of life improved to a point that in developed countries some people had met all of the basic human needs and started to move into the self-actualized stage (as described in "Cultural Development Theory" in chapter 4). This stage is not driven by needs, as the previous ones had been, so people have the opportunity to move beyond being driven. The current trend of human cultural development for the 21st century and the integral age is growing in this direction.

Because of their focus on needs, the cultural development stages through integral can be labeled "tier 1", the "need driven" tier of human evolution[11]. Some may reasonably question that we are culturally still driven by needs in this time of plenty. Looking back at the last century, it started in a time when striving for what was needed was still a real necessity. Life spans were still relatively short and work was hard. World War One emphasized the feelings of struggle. A brief burst of prosperity followed, only to be severely dashed by the great depression. That was followed by World War Two, the Cold War with hot wars in Korea and Viet Nam, and frequent flare-ups elsewhere.

A sense of better times began to manifest in the last quarter of the 20th century. But even then, desires to have it all overpowered the feeling of plenty. From 1985 on, the percent of family income going into savings dropped from a long term average of about 9% to zero and below. We became a nation deep in debt. That drove up stress levels and a continuing sense of need.

Although the average American lives a richer life than the most powerful and the richest kings of two hundred years ago, they feel they need more. Thus, even in a time of plenty, there is a cultural psychology of need. Even some multi-millionaires and billionaires push hard for more, being driven by the psychology of need.

The integral stage appears to be the transition to tier 2, where needs may be generally satisfied allowing for a deeper participation in life. Tier 1 can be thought of as the childhood of mankind while tier 2 can be thought of as a condition of the human race in maturity. With tier 2, a basic change occurs in which a new approach to life becomes available to humanity. With this change comes tremendous possibility as well as responsibility. The first stage of tier 2 (the next stage after integral) has been labeled "post integral" or "holistic".

While the integral stage of cultural development has not become mainstream yet it is emerging now in its early phase of development. One of the first people to talk about the integral stage was philosopher Sri Aurobindo in the early 1900s[12]. He was followed by, among others, philosopher Pierre Teilhard De Chardin mid-1900s. In the last several decades, a growing number of philosophers and scientist have been expanding on the early work.

After a few individuals here and there started living with the integral world view and values, small groups formed and organizations followed. As of early 2012, data indicated about 5% of the American public was moving into integral. Live-online seminars on integral subjects have attracted as many as 50,000 participates from 126 countries[13]. Scientists researching cultural development and world views have had plenty to study in order to map the characteristics of this approaching stage. It is described below.

The Next Stage

The next stage has been tentatively named "integral" cultural development and, like all before, brings a new world view and values into our culture. It has been named "integral" because it not only integrates all previous levels into one, but all cultures and points of view. This stage values each level for its contribution and includes them in its world view. The integral stage is the first to value previous levels of development, which helps to reduce the "culture wars" that have often occurred and are ongoing between earlier levels.

This stage includes a sense of responsibility for the world's problems. In the terms of integral theory it is all inclusive, including all views and cultures. It embraces a natural symbiotic cooperative relationship with fellow humans and nature (our life support system). This view, along with increased world wide networking and flow of creative ideas, is leading to an increased rate of change and camaraderie of the world's population.

Integral is non-patriarchal (as is postmodern), which helps to unleash the creative power of half the world's population. It comes with an increased sense of compassion and respect for all people while maintaining a balance with aspects of realism.

The integral stage embraces the big picture view of evolution and that we are part of it. With that understanding, the opportunity is provided to intelligently plan and participate in evolutionary progress instead of struggling against it and evolving in a random way as previously done. That not only can ease stress but can substantially improve efficiency while reducing evolutionary dead ends. For example, bacteria evolve by producing large numbers of mutations in order to find ones that work. An intelligent purposeful approach can produce a few well thought out deliberate changes.

Integral people do not see themselves as better or as having the only correct view. They recognize that there will always be another stage at a higher order of development, complexity and capability ahead, while learning from and embracing the previous levels.

The emerging stage is not satisfied with business as usual, the status quo, the "system" or the "establishment" (homeostasis), which are too exclusive, limiting and controlling. Those in the emerging stage are change agents for fairness and participation for all. As change agents, they are more like a catalyst, making change happen without being consumed by the process. No longer driven by the hierarchy of needs, they experience much less stress then the current society norm, which is driven at a frantic pace. This new stage will accept nothing less than true global democracy and a world that works for everyone.

Right Where We Should Be

When we look at the worsening global problems of chapter one, we can easily become gloomy about our future. However, as we look at how evolution works, we can see that we are right where we should be.

Remembering the cycles of life of chapter 6, after a period of growth, eventually stagnation occurs followed by decay. In order to transform to a higher level of complexity that can restart the growth process, we must enter the chaos phase where major reorganization takes place. We are at that point now and reorganization is occurring. It means more change than normal and that makes us nervous. But this change is not only beneficial it is required to continue a healthy civilization (a very complex system). In other words, we are at the point where we have entered the decay mode and new growth is starting, and it comes with growing pains. Signs that the old way of doing things are not working anymore are everywhere.

Once we have passed the bifurcation point there is a tendency for culture to follow one of two paths. One is to reorganize to a greater complexity (integral in this case) and the other is to retreat into the past (fundamentalism). Evolution favors the path to greater complexity.

If all this were not happening now, we would be in deep trouble. It would mean business as usual until it would be too late to prevent the collapse of civilization. Fortunately, things are bad enough now to get our attention and push us to change in time. It is easy for most to see that the old order is just not working anymore. Something needs to change. The problem is it is not clear what change is needed, and this uncertainty makes us uncomfortable. This is where looking at the characteristics of integral cultural development and the trends of evolution can help.

The Change is Now

Global problems, like those in chapter 1, require global solutions, applied on a global scale. It is the first time in human history this has been the case and our situation is driving us to a new level of thinking. The stress and chaos are causing our cultural systems to move into reorganization, and is pushing our culture to the higher order of complexity, cooperation and connectivity needed to accommodate the conditions. When the reorganization is complete, a new period of cultural growth will be launched. This is the process we are seeing in its early stages now, and by the time you read this, now may be past (no longer in the early stages). The characteristic of the integral stage, recognizing the good in all other stages, allows for symbiotic cooperative relationships with others in the global community.

On a historic time scale indications are that the integral stage change will be explosive in speed. But on an individual life time scale, change will happen slowly enough that one may not notice it specifically. Changes can easily be missed as they show up day to day, few of them at a significant enough level to be noticed individually as part of a bigger cultural shift. In retrospect, change is seen in an overall context and understood for what it was. Because of our newer understanding of evolution and the fields of science that add insight to how change works, we now have some ability to watch it happen. A full understanding of where evolution is headed will still take a post change analysis.

With the understanding drawn from the long view of evolution and the processes of change, the day to day events of life can take on added significance. What would seem just an interesting, or otherwise unimportant activity, may fit in the bigger picture of cultural development change. For example, if someone came across the information above on the rapid growth of online integral seminar attendance, it would normally be just an unimportant piece of data. But taken in a bigger perspective, the numbers are a sign of the evolving times. Added to other such signs, they start to mean something.

When bacteria are challenged by significant environmental stress, they switch on a high rate of mutations in order to find a fix that works. In other words, they start experimenting in alternate methods through genetic variation. As advanced life forms, humans do this through cultural experimentation rather than genetics.

Cultural experimentation is now happening in many ways by many groups, both large and small, as well as individuals around the world. One example follows.

The animosity between Israelis, Palestinians and Jordanians is well known. In spite of this, health practitioners of these groups have joined into a successful project with a common database to help identify and reduce the threat of infectious diseases[14]. This consortium helped replicate that process in several Southeast Asian countries with similar hostile relationships.

This is just one example of a great many experiments ongoing in the world today. They are cultural experimentations in a manner similar to the bacteria's genetic mutations during times of great stress. While many cultural experiments will fail or not be of significant consequences, some will succeed and result in major change. The point is it there are cultural experiments now in action. Other examples are in chapter 12 and many more can be found in the books *Abundance The Future is Better Than You Think* and *The Rational Optimist: How Prosperity Evolves*[15, 16].

The important point is that widespread experimentation is occurring. While there have always been cases of cultural experimentation throughout human history, the number occurring now has been increasing rapidly just as so much else has speeded up. This is a sign of the cultural change. We will examine these trends of increasing rates of change in the next chapter.

Part IV

A Change of Pace

10

America is a country that doesn't know where it is going but is determined to set a speed record getting there.
–Dr. Laurence J. Peter

The Accelerating Pace of Change

Almost everyone would agree that the pace of change has been speeding up in recent times. Understanding the full significance of this and its next step may, however, surprise many. In this chapter, we will look at the trends of accelerating change and their significance.

Faster Faster Faster
Below we will look at a number of factors that represent the pace of change and their trends over time. If numbers and figures are not your thing, read the words and just glance at the plots, you don't need to understand the details. The figures are plots of data. The numbers themselves are not particularly important. The purpose of the plots is to show trends. Some data, like stock market prices, are irregular and unpredictable. What is presented here, though, has a regular shape curve and can be used to understand general trends. The first is world population growth. This was examined in the first chapter, but it is repeated here for convenience. See Figure 10.1.

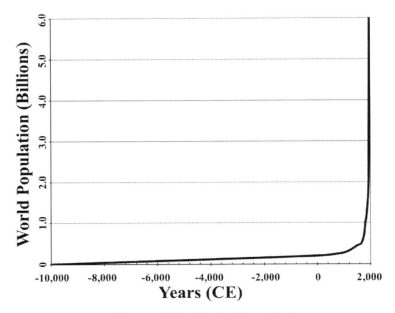

Figure 10.1. World Population

It took from the dawn of humans until 1804 to reach a billion people on earth. It then took only until 1927 (123 years) to reach the second billion. In the eighty-five years between 1927 and 2012, another five billion have been added. The point of explosive growth was reached in the mid 1900's[1]. The population was six billion in 1999 and if it continues at its historic growth rates, it will double to twelve billion by about 2050 and double again to twenty-four billion by about 2100.

The next factor we examine is the rate of cultural development. The data was presented in the last chapter in Table 9.1, and is plotted here in Figure 10.2. The increasing rate of change is quite evident. Again, the pace of growth was rather minimal until recent times, and then it increased explosively.

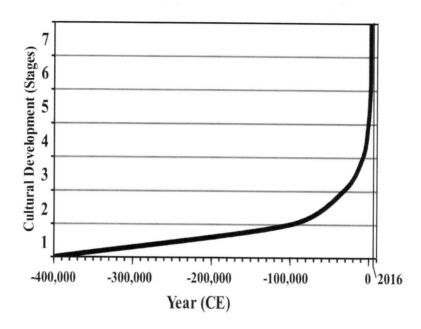

Figure 10.2. Cultural Development Stages over Time

Another factor is the rate of change of technology. There are many ways to measure this. We will show two here. One is the number of major inventions per year[2]. Figure 10.3. The other is to examine the growth in computer power, done here by using Moore's Law, the number of transistors on a microprocessor[3]. Figure 10.4. The current rapid pace of growth in technology is perhaps the easiest for people today to appreciate as it shows up all around us almost daily. Numerous other measures produce similar results.

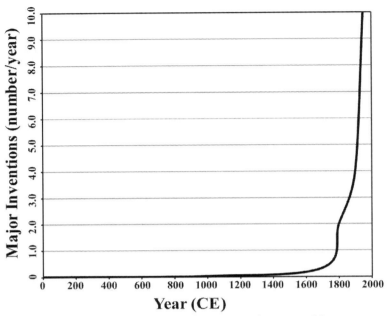

Figure 10.3. Major Inventions per Year

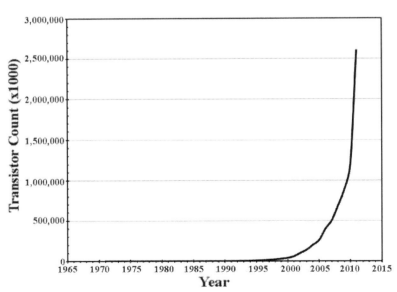

Figure 10.4. Transistor Count on Microprocessors per Year

An additional factor is in the category of information. The flow of information is a measure of our cultural knowledge. It is hard to measure how much knowledge there is in the world at any given time. One way to get a relative measurement is the computer information storage capacity, Figure 10.5, which reached the explosive pace near the end of the 20th century[4]. An estimate of total human knowledge over time has been made by Georges Anderla, who used year 1 CE as the base line and counted all the scientific facts known at that time[5]. That became one unit of knowledge, and then estimates for more current dates were made for comparisons. Figure 10.6.

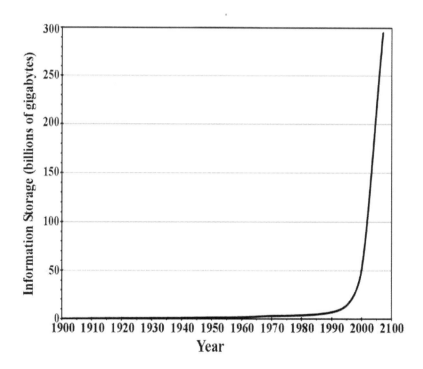

Figure 10.5. Computer Information Storage Capacity over Time

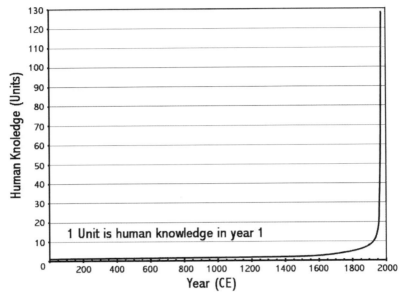

Figure 10.6. Total World Knowledge over Time

Another factor is complexity. To examine complexity we have two measures. One is the amount of internet traffic[6]. Figure 10.7. The internet trend is rather recent but has already reached explosive growth. The other is a method based on the observation that increased complexity requires increased energy flow to maintain[7]. The measurement is based on how much energy flows through a given amount of mass per unit of time. More complex systems have higher energy flow rates. See figure 10.8. Energy flow rates did not start to rise rapidly until after life started on earth, rather late on the universal time scale, thus the late abrupt rise in figure 10.8.

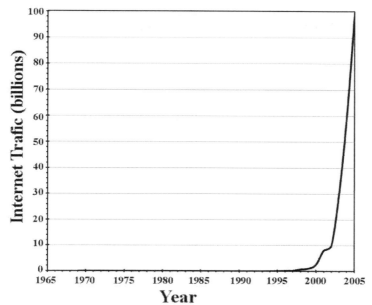
Figure 10.7. Internet Traffic over Time

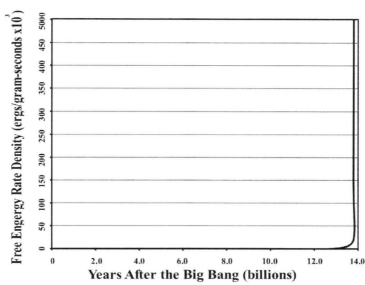
Figure 10.8. Free Energy Flow in Advanced Systems

By now you may be getting tired of seeing plot after plot that look about the same. That is the point: they have a similar shape. Here we are examining factors involved with the pace of life, evolution and cultural development. To be fair, there are many other things we could examine that would not have the same shape curve. But those data are not pertinent to this story. There are many other data sets with a similar shape to the ones above, but we will not belabor the point further[8, 9].

The name for the shape of these curves is "**exponential**". We will not get into the mathematical details of what this means, but several characteristics these types of curves are important to examine.

Exponential Curves

Don't be intimidated by the technical name. Exponential curves are quite normal and common. A simple definition is a curve of data where the next value is larger by a given percentage. Compounded interest of savings is an example. Another characteristic of exponential curves worth understanding well is their nature to rise modestly for some time before they start to take off. That point is sometimes called the "knee" of the curve. There are two stories that are useful in helping illustrate this.

The first story takes place a long time ago when a person presented an impressive chess board he had made to his king. The king was appreciative of the work of art and asked what the subject would like in return. The subject asked for one grain of rice for the first square on the board, then each day for double the previous day's amount until all 64 squares were accounted for. The king agreed thinking that was a trifle. By the sixteenth day, the rice amounted to about a cup worth. Soon it began to increase rather rapidly, and easily exceeded all the rice on earth well before the 64th day. It was impossible for the king to keep his bargain.

Another story is about children, who were told there is one water lily in their pond, but water lilies grow very fast and if allowed to fill the entire pond, would kill all the other living things in it. The children were also told that water lilies doubled in number every day and if left unattended would fill the pond in 30 days, and so should be cut back before then. The children checked on the pond each day. By day 23 less than one percent of its surface was covered, and they concluded there was no need to rush doing the clearing. Less than a week later, on day 28, they noted water lilies covering twenty-five percent of the pond. They started trimming them back, but they needed to work from a small boat and it was slow work. They could cover ten percent of the pond a day. But by then that was not fast enough to keep up with the lily growth. By day 33, the pond was 100 percent covered even though the children worked hard each of the last several days. Figure 10.9 is the plot of the lily growth if left unattended. It shows the now familiar exponential curve.

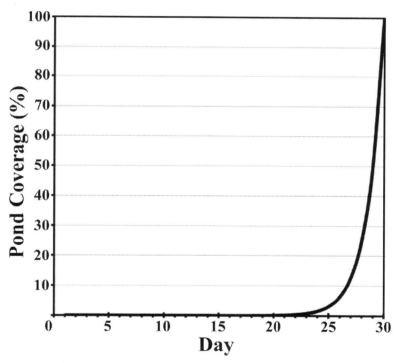

Figure 10.9. The Water Lily Pond Coverage Example

The stories help emphasize one fact too often not appreciated: The scope of change seems more or less normal until all of a sudden it's not. By then things are getting out of hand quickly and if decisive action is not taken, it can be too late. Large organizations, like governments, normally have trouble taking quick and decisive action, especially when such activities ruffle the status quo. Normal reaction to a problem is to put off uncomfortable action until later.

What the last example shows is the danger in ignoring rapidly rising exponential curves, like the ones we have seen above and the problems discussed in chapter one. With the water lilies showing on day 23, not much change was being observed and it was easy to think that's what the near future would be like. But if we are past the knee of an exponential curve, in the explosive growth portion, we may be in for quite a surprise. In fact, it may be too late to start taking corrective actions.

With exponential change, especially in many categories at the same time, it is difficult to make rational predictions about where things will be not far down the road. Which category moves faster and which relatively slower depends on small details in the curves, details too small to see in the broad data that usually goes into the curves. One category may explode a few years before another and the mix makes a big difference in the details of chaos. The only thing that can be said with any certainty is big change is now inevitable. Nothing in human history prepares humanity for what is coming; in fact, just the opposite is the case. Our past is blinding us to the enormity of the present and we will generally be in denial as the multiple categories rush explosively into a future that is unprecedented in historic times.

What exponential curves suggest is that change will keep getting faster and faster until everything is moving at astronomical speeds. No one would be able to keep track of where we were because it would be out of date as soon as it had been noted. Can you imagine a world where reading the morning paper would be a waste of time because by the time you read it, it was too far out of date to bother with? Even checking the news on line would be old news by then. You would have to listen to a live broadcast from reporters on the scene to have it be news rather than out of date already. And reporters on the scene can only say what they are experiencing themselves at that moment and not about what is actually occurring because they only see a part of what is happening. In other words, you could never really know where things were because they would be moving too fast to figure out, and why bother trying when it's already history and something else is in the wind. The wind would be more like a category-five hurricane. Reality would be as stable as a 10.0 scale earthquake. Is this really where it's going, constant change insanity? In the next chapter we will try to put some sense of balance back into the future.

11

You can't stop the future. You can't rewind the past. The only way to learn the secret...is to press play. —Jay Asher

Where We Are Going

We left the last chapter with trends of an increasing pace of change approaching infinity and maybe with some difficulty in believing that could be real. Too many trends say that it is real, yet that seems counterintuitive. In order to better understand these exponential trends, we will take a look at nature and what it can tell us about such things. We are looking for things that happened in the past that might shed some light on what to expect. We need to discuss curves again for a short bit in order to better understand the trends.

Past Exponential Events

Bacteria are the most successful life forms on Earth. They have been around for four billion years. They grow just about everywhere, on the ground, in the ground, underground, in water, in the air, and our bodies contain great quantities of them. We need them to survive. They do not need us (except for some of the ones in our bodies).

If you place some bacteria in a sealed container that has nutrients and track the number of bacteria as they grow and plot the results, you get an exponential curve[1]. But that cannot continue indefinitely because the nutrients are limited. Later, the exponential curve straightens out, starts to slow and then stops growing. Figure 11.1 shows a curve depicting this.

The first part of the curve looks like an exponential curve, but at a point it changes into a decreasing rate of growth (where time = 3 in the figure). That point is referred to as an **inflection point**. The general shape of the curve in Figure 11.1 is similar to the letter "S" and is sometimes referred to as an **S-curve**[2]. Eventually, when the nutrients are exhausted, the bacteria start dying and the population declines. With humans this would be called a famine.

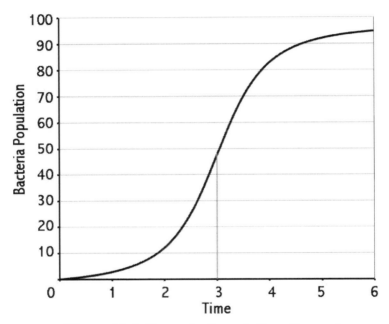

Figure 11.1. Bacteria Growth Rate Example

In the history of bacterial evolution there have been S-curve events. Three of those had major significances not only to life on earth today, but to today's balance of weather, temperature, atmosphere and biosphere. Early bacteria used the chemical soup created during the early days of earth for food. It was plentiful and freely available. Bacteria flourished until they began to diminish the availability of the chemical soup[3]. The first food shortage, about 3.7 billion years ago, eventually forced bacteria to evolve into forms that could make their food by using hydrogen. They did it through a processes like photosynthesis. This was a classical S-curve cycle of growth.

About 3.6 billion years ago, a scarcity of free hydrogen developed and a new food crisis was encountered[4]. Bacteria had to evolve again, this time using carbon-dioxide to produce food and making oxygen as a waste product. The process followed another S-curve. When too much oxygen gathered in the atmosphere, it reached toxic levels and this third crisis resulted in another S-curve of growth as bacteria went through another evolutionary change in order to cope[5].

Another type of curve that occurs in nature is a sine wave curve. If you take an S-curve, then add another one upside down after it and repeat a number of times, you get a sine wave. It resembles wind-blown ripples on a lake or ocean swells. Figure 11.2.

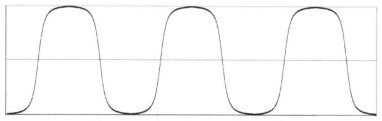

Figure 11.2. A Typical Sine Wave

Animal populations tend to grow and shrink in patterns generally like sine waves, although with considerable variability, not as neat and even as the curve shown. When food is plentiful and predators limited, the populations start a rapid growth. As the population grows, overgrazing leads to food shortages and at the same time, predator populations are increasing. Eventually a famine sets in and population numbers dive. The cycle repeats over and over. There is the initial exponential shape of rapid growth, followed by an inflection point, then an end to growth as in an S-curve, and then the decline which forms one cycle of a sine wave.

There is one more type of curve to look at, a curve that has what is called a "discontinuity" in it, sometimes called a "singularity". In Figure 11.3, for a value of x = 20, there is no mathematical value for y.

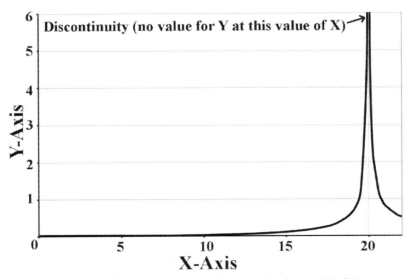

Figure 11.3. Curve with a Discontinuity at X=20.

The curve just stops at that point, then continues on after when x is larger than 20. It's also called a singularity because it occurs at a single value of x only where there is no mathematical solution. There are few singularities in nature, but they do occur. Examples include black holes, the moment of the Big Bang, and in the evolutionary progress of massive stars.

A plot of the different stages of thermonuclear fusion vs. time in a massive star has a shape like figure 11.3. The burning of fuel accelerates rapidly at the end and results in a singularity. That event is called a "**supernova**". It's a real blast, and it happens in a flash of time. At that moment, the star undergoes a massive explosion and either a black hole or super dense neutron star results. This is called a "phase change" or a "change of state".

If we look at the chemical H_2O, we call it ice when it is below 0^0 Centigrade, water when it is between 0^0 and 100^0, and steam when it is above 100^0 Centigrade. These are three states or phases of H_2O. If we take a certain amount of ice (say at -50^0 C), for each calorie of heat we add, the temperature will go up 1^0 C. This continues until the temperature reaches 0^0 C, at which point the temperature stops changing when we add heat. The ice will be melting. When it is all melted, the temperature will start to rise as before with each addition of heat. The period when the ice is changing to water is called a "state change". Another state change occurs at 100^0 C when the water turns into steam.

Now that we have examined some examples of curves in nature, the question is which one seems to fit our current situation? In other words, what can we say about the likely trend we will see in our near future? The continued exponential curve of explosive growth we are currently seeing cannot continue for the long run for the same reason bacterial growth changes into an S-curve – limited resources. Sine waves, where there is cyclical growth in shorter time periods but no net growth over long intervals is not the way evolution has worked during the 13.8 billion years since the Big Bang, so it is not likely to be what is going to happen next. While singularities do occasionally occur in nature, they have not been a normal part of evolution's increasing complexity process. A sudden complete change of human life into another radically different form seems unlikely, while a sudden stop to change due to some kind of a change in state does not appear to be a practical answer either.

That leaves us with an S-curve as the likely trend. S-curves describe processes that are found throughout nature. They can form chains, end to end in a cascading growth. In fact, S-curves are fractal in nature, often with a series of smaller ones imbedded in large curves[6]. A well-known adage in science is to always use the simplest explanation. More complex explanations offer many more opportunities for error. The S-curve is the simplest explanation for the current exponential-like trends. Each stage of cultural development has followed an S-curve during its development. It makes sense that the larger shape of the curve of cultural development is also an S-curve.

With the preponderance of natural examples of S-curves and the previous cultural development examples, it is assumed here that we can reasonably expect we are also on an S-curve rather than an unending exponential or other type curve. Why all this talk about what kind of curve? It makes a tremendous difference when looking into the near future as to which trend we project. By following known examples from nature and past evolutionary trends, we are on far firmer ground than picking something based on a radically different trend. Before we venture into the future, let's revisit the population data.

Recent Changes

In the last chapter we looked at the world population over that last 12,000 years in figure 10.1. It looks like an exponential curve. However, if we examine a shorter time scale in closer detail, the curve stops increasing in slope (getting stepper) and starts to decrease its steepness in the mid-1980's, figure 11.4. The curve has reached an inflection point, where the steepness stops increasing, reversing the accelerating trend[7]. This is one indication that the S-curve trend is the correct choice. Substantial change has been occurring in family size in some of the most populous countries such as China and India. These trends are likely to continue. The United Nations' projection indicates that the population will continue to grow but at a slowing rate[7]. Figure 11.5.

In the last couple of decades, human cultural complexity has been rising at explosive rates. That includes just about everything human, from technology, the global community and ordinary living. The global infrastructure, by way of trading, business, travel and communication has grown just as fast.

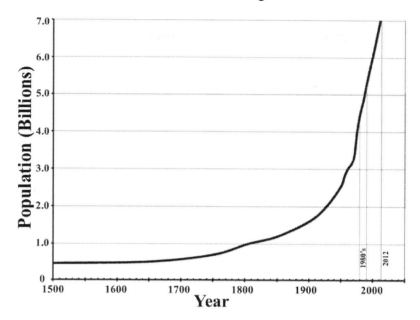

Figure 11.4. World Population

With internet traffic rate having skyrocketed in about 2002, cell phone use having grown to over 50% of the world population, and smart phone sales approaching a billion per year, the world is interconnected in a way few had dreamed of 50 years ago.

This interconnectivity is the equivalent of a global central nervous system and our world is beginning to throb to a common beat. Ideas and innovation are contagious and instantaneous. The old limitations of location, local culture and country are slipping into the past as a global human culture emerges. In the past, a new fresh creative idea may never be heard of in many parts of the world. Today, it is the topic of conversation in village meeting places in far corners of the world hours after being posted on the net.

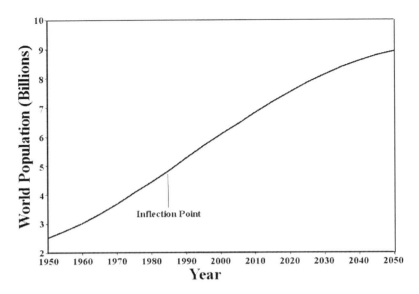

Figure 11.5. United Nations World Population Projections

Likely Trends in Near Future

Table 9.1 shows each time interval between stages of cultural development has been about an order of magnitude shorter than the one preceding it. The integral stage is beginning to burst forth now. If integral goes mainstream in a few years, as trends suggest, that would mean that the last time-interval between stages (postmodern to integral) would be about 50 years. Extrapolating to the holistic stage gives 5 to 10 years after integral starts or in the range of 10 to 20 years from now. That would mean, even as the Integral stage is growing in significance, the holistic stage will be following right on its heels. Indeed, there are signs this may be so. In effect, this would be like two stages merged into one big double whammy. That would be like riding on a world class roller coaster, starting off with the bottom dropping out from under us, in a manner of speaking. On an exponential curve, that would put a post-holistic stage only 2 to 5 years thereafter. But if we are actually on an S-curve trend, at least there is bound to be an inflection point somewhere about then. Evidence suggests that five years is about the minimum time it takes for an individual to make the move from one stage of cultural development to the next.

Trying to predict where the inflection point is risky. However, there is a point that tends to fit naturally. Cultural development stages 1 through 7 are all tier 1 (discussed in chapter 9), driven by needs (and their fears). Stage 8, Holistic, is showing signs of being different enough that it is called tier 2. It would seem that the inflection point would also be at the junction of tier 1 and 2, between Integral and Holistic. This also fits the indication that the interval trend between integral and holistic is about as short as possible, requiring a stop in acceleration. If so, the post-holistic stage would be more like 50 years after the holistic stage rather than the 2 to 5 years of the exponential curve. An S-curve is much more realistic than almost instant change. Thereafter, the pace of change may slow. This would be fitting because the changes in tier 2 will likely bring huge changes to human cultures in every aspect. Again, it should be cautioned that this level of extrapolation of trends is getting into the speculative range.

Because the next two stages will probably overlap so closely, we will examine the characteristics of the integral and holistic stages together to project some of the effects they are likely to have on our culture. These characteristics will not suddenly become the norm. They will, as in past stage changes, start to grow in influence within our overall culture and take time to have a major effect. This time the change will happen much faster than any previous one due to the general quickness of today's pace.

Perhaps the most notable integral/holistic change will be a global attitude, as compared to current national and regional attitudes. Countries will work more closely together and planning will be on a global scale. Problems will be handled on a global rather than national level, even when affecting a local area. This trend is already starting to happen, with limited success so far.

Politicians will start to respect others' views and work together rather than living in perpetual conflict and constantly trying to destroy each other. Campaigning will become more based on issues, not mudslinging and image making. People will demand it.

Global cooperative efforts on saving our life support system, the environment, will rapidly grow and expand. Sustainability will be included in these efforts. World poverty, slums, education and health will be tackled in an effective way. Competition will be balanced by a sense of fairness and cooperation. All people will be respected and honored. Individuals will become less reactive to ego impulse and will learn to evaluate ego impulses before automatically responding.

As these influences grow in strength, conflict, war and strife will subside. As global poverty, slums, ignorance and disease decline, the sense of hopelessness they inspire will also decline. Terrorism will follow suit.

At this point, it is likely that many readers will be thinking something like: "this is a utopian dream and not real". In fact, it is not real now. As each of us reads, we are interpreting though our own world views and sense of reality. Unless we are at the integral or holistic stage, we will not be able to easily accept the broader world view yet to come. This is where examining trends can help us to open our belief ability to consider what can be before automatically rejecting the broader view. Throughout time humans have had difficulty in seeing into the future because they generally were looking through their current world view, which is based on their past experience. For most of history, change was slow enough that that method worked reasonably well. But in exponential times, past experience fails badly as a way to think about the future.

Figure 11.6 shows a linear extrapolation based on the past as compared to where the future goes in an exponential fashion. The difference between what we expect and what happens can be enormous.

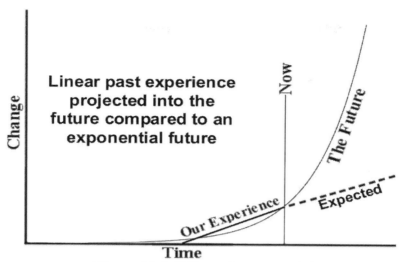

Figure 11.6. Effect of Exponential Curve on our Expectations

Understanding these cultural development stages and being able to extrapolate into the near future is a new and still developing science. It gives us, for the first time, the ability to understand some of the likely trends. Because there are already some people at these higher development levels, small as they are in percentage, they still offer enough evidence to support a reasonable extrapolation.

We are currently in the phase of transformation called chaos. The details of how all these changes will come about are not predictable, but the reorganization of world cultures follows the "strange attractors" of chaos. Those are the broad extrapolations made in this chapter. Evolution continues to move forward and our current world views and realities are guaranteed to change, and so is the world we live in.

Changes in "Less Developed" Countries

Cultural development has been somewhat slower in the less developed countries compared to the United States and other "developed" countries. In most of the less developed countries the majority of people are still at the traditional stage. It could be argued that if it took over 350 years for the developed countries to transition into postmodern, it should take that long for the less developed countries. There is one massively significant difference though.

When the developed countries made the move, it was new in the world and no one knew what to expect or where it could go, or even if it would work. In the last couple of decade's two things alone changed all that in the underdeveloped countries; the internet and cell phones. Today, information and ideas are at the speed of light. Even in small remote communities in Africa, Asia and South America, ideas like integral philosophy are being discussed. Topics like sustainability and conservation are being discussed.

Today, those in the remote areas of Africa can have more communicating power in their hands than anyone anywhere did 15 years ago. The change going on there is based on existing ideas and ideals that are at work in developed countries. What is done in the developed countries is learned in the less developed countries immediately. The time lag of decades or centuries is rapidly vanishing.

We tend to look at the future with a linear prospective while it is actually developing exponentially at this time. What happened in the past is not how it will happen today or tomorrow. With the full flow of ideas and the attraction of the success of the developed countries, the less developed are developing at speeds never before seen. This is happening today throughout the world. Many third world youth of today already sees a future unimagined by their parents.

Effects of Technological Changes

The changes in cultural development will lead to new ways of thinking providing opportunities to solve many of our current problems. It is the combined effect of new ways in thinking together with new technologies that promise to solve the rest of our global problems. While it seems likely that the S-curve inflection point of cultural development will occur at about the integral/holistic point, or roughly in 5 to 20 years' time, the inflection point for the explosive technology growth is not necessarily at the same time. Technology however is a product of human action and if human action follows an S-curve, then technology likely will follow with an inflection point somewhat later.

Those familiar with the latest advanced technological work in progress, and thinking in an exponential way, are seeing some truly fantastic things coming, some very soon, and others a little ways off. Technology is moving so fast, it is hardly worth listing any of them here, as by the time you read this, they may be current news. Nevertheless, several follow below.

While 3D printing is still fairly new, there are already 3D printing manufacturing shops in production, including printing parts for new 3D printers[8]. This technology promises to revolutionize manufacturing.

Developers are already experimenting on inducing stem cell production from a person's own skin, then using these cells in a 3D printer to print replacement body parts[9].

Developers have already printed a mini experimental kidney and are thinking they could make real kidneys for transplants in about a decade. It's not quite like the Star Trek machine that could make anything you ordered, but it's getting much closer to that reality. Simple 3D printers are now on the market for much less than the cost of a dot-matrix impact printer of 25 years ago (if you even know what a dot-matrix printer was).

Computers will be going quantum one day not far off. An early version is currently being developed for large application production. The possibilities of what quantum computers would be able to do are still being thought out, but the potential is for computing to go so far beyond today's capabilities as to make comparisons difficult at this time.

Other areas, each with revolutionary advances likely, include energy production, chemistry, nanotechnology, micro machines and micro manufacturing, robotics, and health care. We are on the brink of so much advancement in capabilities, that it is almost too much for any one person to comprehend. It would take months of detailed study to even come close to grasping the magnitude of technological change on the horizon.

We are on track for revolutionary/evolutionary changes in cultural development leading to the maturity needed to handle the revolutionary changes technology is bringing. While it is too early to see how this will all come together, all the pieces are in place or on the horizon to bring an abundant, sustainable, cooperative and peaceful world for all into manifestation. You have the unique distinction of being alive to witness the first really major transformation of the human race since becoming Homo sapiens (from tier 1 to tier 2 and a double cultural stage change in one life time). This is big stuff. It's transformation with a capital "T". The bad news is we do not have any choice about it; it's happening. The good news is, at least now we have some idea what is going on. If we think there must be a better way for humanity to live, there is, and this is how we get there.

These are exciting times. They are also very trying times as we witness the death of the old way of being human, struggle to find the new, and deal with the unsettling lack of certainty. Take heart: It's time to mourn the death of the old and participate in the birth of the new.

In our next chapter we look at some ways that the developing technological advances can solve some of our global problems when used by a global, one world culture.

Part V

Where Do We Go From Here?

12

The present convergence of crises—in money, energy, education, health, water, soil, climate, politics, the environment, and more—is a birth crisis, expelling us from the old world into a new. We sense that 'normal' isn't coming back, that we are being born into a new normal: a new kind of society, a new relationship to the earth, a new experience of being human. —Charles Eisenstein

The Promise of Now

Given the scope and complexity of the problems described in chapter one, it is easy to doubt that technology can save the day. Even though we may understand exponential trends, we tend to feel through linear thinking and therefore may not feel good about our future prospects.

The technologies we are talking about are not just another step in normal progress, such as a next generation smart phone. They each are revolutionary in that they are not just improvements but will change the basic way things are done.

Likewise, the cultural developmental changes are also on the scale of revolutionary. To put it into perspective, in the developed countries during the next couple of decades we may experience as much cultural developmental change as in the past 400 years. In the less developed countries, it may take an extra decade or so for this change, but the magnitude of change may be as much for them as the last millennium was for the more developed cultures.

All that change is hard for our minds to grasp. Our normal reaction is "No, I don't think so". Figuratively speaking, that reaction is the mountain blocking our view of tomorrow. But tomorrow is coming and fast. The dawn is already breaking. We have passed the "knee" of the exponential curve and are in the explosive change stage. Our thinking is still influenced by what is considered normal, which is the past, when the curve was still slower. The rest of this chapter is about tomorrow. It's beginning to show itself through technologies that are either currently in development, in advanced research stages, or already in the early stages of implementation. The following will be a few examples of what is happening and can happen. Many more examples exist but it would fill another book[1].

New Style of Philanthropy

To solve some of the global problems, it would help if solutions were not limited by government funding, red tape, special interests and bureaucracy. Such an effort would undoubtedly need large sums of money and the creative talent for implementing new ideas and doing the impossible. Fortunately, a new breed of philanthropists is offering such an option.

With the computer, internet and related electronics, a new generation of multi-billionaires has evolved in the last several decades. In past times, the super-rich often gave money for such things as libraries, concert halls, and hospital wings, mostly in their own area of the country and named after themselves. There has been a major change in recent years.

There is a new breed of "dot-com" philanthropists. Many are globally minded and interested in all people. They have made their billions by innovate out-of-the box thinking, not stopped by "you can't do that" thinking and not afraid of risk or failure. Government programs are run by bureaucrats who can't afford failure and therefore take the slow and safe way. The "dot-com" philanthropists have a history of pushing boundaries, doing the impossible, taking risks and sometimes failing. But they know failure is just a lesson and by learning what does not work, they know better what does. They have one other significant asset; billions of dollars.

Examples of this new breed are Bill and Melinda Gates and Warren Buffett. As previously noted, they have set aside a vast fortune of many billions of dollars to help the world's people. There is a major difference between older ideas of giving money to charities that do good work, and the new philanthropy. A charity could, for example, use some money each year to feed a group of impoverished people. The new philanthropist uses money to leverage its effect, such as teaching the impoverished to grow food and helping them with what that takes. Even more, they teach how to expand and make a growing business out of it so other people can be fed. In time, the people need no more help to thrive. (With old style charities, if the money stops, the handouts stop and people starve.) With the new philanthropist, it's about empowering people to unleash their creativity and resourcefulness to help themselves and others as well. It should be noted that some charitable organizations are also taking this approach now.

The new breed does one more thing, they monitor and stay in touch with what's going on. If a blockage occurs they add horsepower to move past it. They invest in activities that promise to pay dividends in productivity down the road, not just in giving handouts. Some of the programs not only look for the financial benefits of new businesses, they have measurable environmental or social goals to meet as well.

Many of these philanthropists get involved with the people in remote places around the world. Imagine Bill Gates maneuvering the mammoth Microsoft Corporation at one point, then in a small remote country getting dirty helping to plant a crop. For these philanthropists, it's not about fame, boasting rights or showing off, it's about making a difference in the world.

Some philanthropists fund advanced research on ideas that can lead to revolutionary advances in solving global problems. By careful use of seed money, innovations that may not have otherwise come along until much later can be jump started producing significant benefits much sooner.

Feeding a Hungry World

Today, hunger affects about one in seven people in the world, or approximately one-billion[1]. There is enough food to feed them all now. There are many reasons it's not being done, including costs, politics, local conflicts and limited infrastructures.

The bigger problem is how we feed a world population when it doubles to fourteen-billion as it is expected to by mid-century. Although we may have enough food to feed our current seven-billion world population, we will need about 28-percent more to feed just nine-billion. At the same time farmable land is diminishing and so is available fresh water for farming.

With global climate change and greater fluctuations in weather, both droughts and floods are becoming more severe, having greater adverse effects on crops. With current methods, it would be difficult just to maintain current food outputs. What is needed are new approaches, which need to be more energy and water efficient and not require massive arable land to implement.

What we really need for the future population is a way of growing food that takes much less land, water, pesticides and energy. A solution has been proposed, tested and put into effect already - hydroponics, a method of growing food in water. Hydroponics was first used in the 1930's and has grown slowly. While normal agriculture uses about 70% of earth's available water supply, hydroponics uses 90% less, which would only be about 7% of all water supply[2]. Hydroponic food first made it into the supermarkets in 1986.

A newer method called aeroponics has been under development to grow plants suspended in air. Water spray or mist with nutrients is used to grow the plants. The process uses only 70% as much water as hydroponics[2]. If all agriculture were converted to aeroponics, it would only require just several percent of the planets available water.

There are other significant aspects to aeroponics. Besides using vastly less water, it can be used in multi-storied buildings, saving land usage, and can be built on non-arable land. The buildings can be close to population centers substantially reducing transportation costs. It also reduces the carbon footprint.

Aeroponics uses less water, less land, less energy, produces no pollutants and can also expand the growing season. Vertical agriculture is already being used in commercial food production in Singapore where they are running out of land[3]. The vegetables are preferred to imports because they are locally grown and fresher.

Vertical aeroponic vegetable gardens may be a big step forward, but what about protein? World livestock production takes an astonishing 30% of the all the agriculture land on the planet. World meat demand is increasing at a rate faster than population. There is mathematically not enough land to keep up with this demand. The energy needed to grow the livestock is over 50 times the energy produced by the protein in the form of food[4]. The future population growth will not sustain such inefficiencies.

Research now being conducted shows a way to make the vertical vegetable garden idea work for protein as well. By producing cultured meat in laboratories and using 3-D printing machines, meat can be produced without first growing animals, butchering and processing. Cultured meat is made in-vitro (using components of an organism that have been isolated from their usual biological surroundings). By growing more of these cells, then using a 3-D printer, a meat can be assembled. There are advantages to this process.

Cultured meat does not have the gastrointestinal tract with its toxins, so there is little chance of contamination as sometimes occurs in butchering living animals. There is also the humanitarian aspect of not having to kill an animal in order to eat it. Also, with butchered animal meat, there are properties that are not healthful to the human heart. With cultured meat, these can be reduced or eliminated, resulting in heart friendly meat. Chemical additives can be avoided.

This process promises the ability to free up vast land areas now dedicated to livestock, and replace them with factory style vertical buildings where cultured meat is produced with substantially less energy and resource requirements. Reductions in pollution would also be realized. As with vegetable growing buildings, these protein factories can be located outside population centers on land not suitable for farming, while reducing transportation costs and carbon footprints.

The first prototype hamburger has already been created as a research project, cooked and eaten (with no side effects)[5]. Although it did not have much flavor, it was a first try and subsequent research efforts will add flavor elements. While not ready for production yet, the idea is producing active research to perfect the processes needed to make it happen. The hamburger was funded by a "dot-com" billionaire. Making 3-D printed hamburger is the next step.

Some readers may be thinking at this point "Yuk". To help put things in perspective, think what someone in the 1850's would say if we described our current food processing. It may well have been "Yuk". When we get used to different things, what one day was a "Yuk" becomes another day's "normal". The question is not whether cultured meat is a good idea today, but the fact that it may be a necessity in order to feed twice our current population. This is the type of revolutionary change that can solve a global problem. Conventionally raised "real" meat could still be produced while prices and convenience encourage many people to switch to more efficiently manufactured meat.

Educating the World's Population

Ultimately, in order to solve the global problems we must deal with poverty, hunger and the population explosion. One of the biggest obstacles is the lack of education. This may seem like an almost insurmountable obstacle if one thinks about it. Think of all the teachers to be trained, all the school buildings to build, the incredible cost and all the political obstacles. But there are other ways now in place and developing.

The education system in America was developed during in industrial revolution and based on the factory methodology. The concept was to pour standard information into children and make them conform to the standard. Testing produced rule followers. The current system does not do well at teaching children to think analytically or critically. Children don't know how to evaluate information or apply what they learned. The mechanical process of rewards for right answers and punishment for wrong answers teaches children not to take risks or be inventive. Many are not prepared for life in the 21st Century. This is not the only way to learn, and probably not the best. Our education has not kept up with modern life.

To understand what is possible in the less developed parts of the world, a story started in India is useful[8]. Sugata Mitra was head of research and development for a technology company in Delhi, India. He had an office in a large modern building on the edge of a slum. A high wall separated the office building from the slum. The children of the slum did not speak English and had no understanding of computers or the internet.

Mitra was curious about what would happen if these children were exposed to computers. In 1999 he built a hole-in-the-wall just big enough to install a computer keyboard with a track ball and monitor facing the slum so that it could not be removed but could be used. The computer was connected to the internet and was always on. He also installed a video camera to monitor what happened. The children were uneducated but they were curious.

Within a very short time they could maneuver the curser and click on things, all without any instruction at all. Within a matter of hours some children were surfing the web, the web they did not even know about earlier in the day. Impressed by the results, Mitra set up another experiment in a remote location well away from New Delhi. There were no people in the slum who knew anything about computers or the internet. He had the same results. Mitra repeated this experiment in a small rural village and again, the same result. This experiment has been repeated in numerous locations in India and Cambodia producing consistent results.

Mitra then enlarged his experiments to find just how much children could learn on their own. He set up one experiment for 12-year-olds who knew nothing of the internet and did not speak English. He left them with the computers and told them there was information in the computers that was very hard to understand and he would be back in two months to test them about it. The subject was biotechnology and it was in English. Biotechnology was a subject none of them had even heard about.

Two months later they understood something about English and biotechnology. They did poorly on test scores compared to well-educated students of the well to do, but the fact that they could do so well in only two months was spectacular progress. Two more months and their scores were as good as top schools in New Delhi. His work inspired the movie *Slumdog Millionaire*.

To move out of poverty, people need education. When they get it, their world changes. Today, organizations like OLPC (One Laptop Per Child) have managed to reduce the price of computers to a small fraction of what they once were, and have delivered millions of them to children around the world. Where these computers have been delivered, high truancy rates drop to zero[6]. The Gates Foundation found that the majority of dropouts were due to boredom and the feeling of the irrelevancy of what was being taught. Computers changed that.

Computers by themselves are only part of a complete education. Salman Khan in 2004 started tutoring a cousin over the internet. He started making videos to do this, and then made the videos public. They became popular.

In 2009 Khan quit his job to work full time on what became the Khan Academy. Continually adding more content and subjects, as of December 2013 his free academy had one and a half million subscribers from around the world and the videos have been viewed over a third of a billion times[7]. Funded in part by billionaires, Khan has added a feature that allows teachers to connect with students in order to monitor their progress.

While not a substitute for a formal education, the Khan Academy opens doors to millions of people around the world who may not have access to a formal education. Today, many

Universities are making online courses available to non-students for not-for-credit use.

One additional aspect has been an effort to create an artificial intelligence computer tutor. Such a program has already been used and helped students substantially increase test scores. The program is able to respond to the level of the students to help each learn better.

The vision of educating the masses of poor in the world is taking a major step forward. Even if only a portion of each population gets exposed to a meaningful education, the effect on the others will be significant. It's the first step to eradicating illiteracy and ignorance.

Providing a Healthy Environment

Every year millions of people die from preventable or treatable diseases. Many millions more are significantly impacted or debilitated by them. Past ways of thinking about disease and their limitations are beginning to transform and the promise of major improvements in global health are growing.

Changing the methods and costs of diagnoses is one step well underway. Using a special printer to print medical test strips that can be used in remote areas is one way to do what used to require a $50,000 machine in a laboratory. Another idea under development is using micro-chips that can take a drop of saliva and determine pathogens for minimal cost. The technique has already been developed that reduces HIV testing from weeks to a few minutes and a few cents using micro-chips[8].

The Bill & Melinda Gates Foundation has been making a significant impact on world health, giving billions of dollars to many efforts. The foundation has focused on reducing or eliminating tuberculosis, malaria, polio and HIV among others. Their efforts include basic research, low cost testing and treatment and infrastructure[9].

The foundation has also undertaken to solve the major sanitation issue for billions of people, that of human waste. The toilet has not significantly changed since it was introduced in the seventeen-hundreds. Efforts have been funded to come up with a modern design that does not require a source of water and a sewer connection or an electrical power connection and yet is cost affordable for the billions of poor who need such a toilet.

Projects such as these are expected to pay off in a substantial benefit to the lives and health of several billion people in the coming years.

One concern some may have is that if we make several billion people's health better, will this not make the population problem much worse. While that could well be argued it is not a good reason to allow suffering when it can be avoided. In the long run this better health may actually help solve the population problem, as we shall see later.

Water

Less than one-percent of the world's water is drinkable. The rest is salt-water or polar ice[10]. That fraction of a percent is in increasingly short supply as the demands of a growing population for drinking, farming and flushing compete.

With the majority of the world's population living near oceans or seas, desalinization of salt-water may seem a good solution. In parts of the world, people are already using desalinization plants for part of their water supply. Because of the high energy needed to run these plants, this is only a good solution for limited applications.

Technology now being developed may solve much of the increasing problem of decreasing drinkable water supply. **Nanotechnology**, technology of the extremely small scale, is forging a number of innovations. One creative idea uses a membrane with pores only 15 nanometers (6×10^{-7} inches) wide. By passing water through it, bacteria, viruses, parasites and fungi are removed, making waste water pure and useable for drinking. A filter container about the size of a 5-gallon can could provide clean water for a family of four for several years at a cost of about three cents per week[10]. Another nanotechnology filter has been tested that can remove salt.

A different method has been developed that distills waste water to produce sterile water and recovers most of the energy used. Any type of liquid that contains some water can be put in and sterile water comes out[10]. It takes a fuel to run, but anything that burns will work. A test version has been running on cow dung in a village in Bangladesh. A byproduct is a modest amount of electricity. The device is designed to run without maintenance for five years or more.

The problem of desalinization of salt-water to satisfy the growing demand for fresh water seems solvable on a global scale in the near future, not only for well developed countries but for the poor in less developed countries.

Energy

Electrical energy use has been increasing at a rapid rate. Not only does electrical generation mostly use non-renewable fossil fuels, it also contributes to air pollution. Many hundreds of millions of people still live without any electricity. Alternate methods of providing electricity are under development.

Photovoltaic (using solar cells) electricity generation is not new. The technology has been around for many decades, but the cells were too expensive to compete with conventional electrical generation in most cases. The efficiencies of solar cells have been increasing while the cost of production has been decreasing. At some point, expected within the next several years, the cost of using solar cells will drop below that for conventional electricity. At that point, many will start to use solar cells for normal home and business needs, in part to reduce pollution and consumption of non-renewable fuels but mainly to save money.

When it becomes a financial plus to install solar cells, it will become big business. The costs of mass production will fall even more, which will make providing electricity to the hundreds of millions in remote areas feasible. A small village could provide solar power for a cell phone tower, a communal satellite TV, and some night lights, all without bringing in expensive power lines.

Another aspect of solar cells is storing power during daylight for use at night. Advances in energy storage are showing the promise of being able to supply substantial energy storage at costs that make solar cells cheaper to use than current electrical power. This means that when there is enough sunlight during the day, the electricity generated could last through the night until the next day, further reducing drain on the electrical grid.

An additional prospect in development could become a major player in world electrical energy generation. Its name is "Traveling Wave Reactor" (TWR). The idea of nuclear power seems horrendous to most people today. After the disasters at Three Mile Island in 1979, Chernobyl in 1986 and Fukushima in 2011, as well as numerous less published incidents, public confidence in nuclear power is almost non-existent for good reason. It's just too dangerous.

Suppose, for the sake of discussion, a new technology could be developed to produce a device that used the byproducts and waste of old nuclear power plants and weapons manufacturing for fuel, and could not suffer a meltdown or explode. Suppose it also produced only small amounts of relatively safe byproducts, had no moving parts, was pollution free and did not require enriching uranium and could run for half a century with no additional fuel or maintenance. Sound like a fantasy dream. While no such device has yet been made, it is under serious development[14]. A number of prominent venture capitalists and the Bill Gates & Melinda Foundation have funded the effort.

TWR power plants, using current stockpiles of waste products from past enriching uranium processes, could provide enough fuel to power all the current energy needs of the world for many centuries. No more non-renewable fossil fuel consumption, with its resultant pollution. Such power plants could be sealed units small enough to be transported to remote areas and provide local power without a regional distribution system.

While no one can guarantee success, TWR has many backers who think success likely. A functioning demonstration unit may be operating by 2020[11]. In theory, its power could be cheaper than coal, while helping to eliminate current nuclear waste.

Eradicating World Poverty

While billons of people live in poverty, social unrest, revolutions and terrorism will continue to flourish. Where there is no hope there is strife. What some of the above efforts demonstrate is that there is now a growing new frontier of education, information, networking, and communication that is starting to make inroads everywhere. As we shift further into a true global cooperative culture, more efforts to provide these benefits to everyone will continue to grow.

When uneducated people, who struggled to live start understanding the bigger world around them and begin to see what's possible, many come alive with new hope and determination. New entrepreneurs come out of the woodwork and start new businesses. A middle class grows where there had not been much of one, and a healthy middle class is recognized as the bedrock of a strong and flourishing economy.

There have been many examples of this already starting to happen in the less developed areas of the world. When people begin to see things improving they gain hope. Where there is education and hope, there is less strife and terrorism. It is in everyone's ultimate interest to support these efforts, even if only for selfish reasons. As the world's poor start moving toward middle class, the explosive growth in business opportunities and trade that will result is exciting to contemplate.

Solving the Population Problem

When the progress mentioned in the above sections has been made, it turns out that evidence suggests the population problem will just vanish. High reproduction rates and poverty go hand in hand. Today, in the more developed countries, the birthrates are low, close to, or even below breakeven (the point where the birthrate maintains a stable size population, not counting immigration).

As childhood mortality rates drop, so does the birthrate[12]. There are specific examples in some countries where there has been a direct inverse relationship between child-mortality rates and direct relationship with life-expectancy compared with population growth. In other words, when life-expectancy increases, population growth slows and when child-mortality rates decrease, so does population growth. Both these effects will likely result from the improvements of on-going programs and others that will surely follow.

While not all the ideas outlined in this chapter may work out, the direction of action is clear. Many other ideas not yet evident will also come to pass. The general trend here is to encourage the less developed portions of the world to become more prosperous while at the same time solving the global problems that currently threaten. The next chapter looks at how to proceed.

13

When the winds of change blow, some people build walls, others build windmills. – Dutch proverb.

Riding the Winds of Change

Some people like to read the last chapter first to see if they want to read the rest of the book. If you are doing that, you will have missed the most important parts and will not really understand this chapter much. Please read the first paragraph of the preface.

Transformational Change

We are in transition now, a big one. It is not only a transformational change, but for humans, an unprecedented transformation. Now is the time to grieve our loss of the old ways and move into the regenerative in-between phase in preparation for the new. We can do this in great fear and increase our stress and therefore the chaos and pain or we can look forward to the rebirth and a better order of human life. We are as if in winter but we can look forward to the new birth and creativity of the approaching spring. It is our attitude that can mitigate the severity of the winter. Understanding that this transformation is a natural process leading to a better future can help immensely in our attitude on a cold winter day when it seems dark and blustery.

We have covered from the big picture view of things, a history of time, space, matter and life. We have examined the common themes and trends and then extrapolated these trends into the future several decades. This extrapolation happens to cross a major step in human evolution, the adding of a new layer of holon. The outcome in detail is unpredictable, but the overall pattern is not. Based in the overall pattern, out of the chaos of change, the strange attractors point to a spectacular future for human kind, or whatever we will call ourselves then (perhaps Homo Sapiens Provectus - *advanced*). This step may be as big as the last new holon layer, evolving from primate to human. The massive change requires considerable new levels of complexity, interconnected networks of communication and feedback loops. We have been seeing much of this developing in the last several decades as well as how the speed of evolution has been accelerating ever since the Big Bang started it all. It is happening so fast, that most people now alive will probably see the next level in their lifetime.

In one sense, this scenario seems so unlikely that it is easy to scoff at. But this is where the overall trends of billions of years are pointing. So often in science the first people to see a new idea and report it are accused of many things, none of which are particularly flattering. Time will tell which is true. If you choose to think this story preposterous, it is. But if you choose to ignore it as nonsense, I ask one thing: keep it in the back corner of your mind and as things change, notice if they go the way indicated here.

With holons, the earlier levels support the later, while the later levels provide additional emergent qualities that do not exist within the earlier. Along with the inner networking, communications and feedback loops of the more complex organization, the relationship between holons is "symbiotic". A symbiotic relationship is one where two or more entities (holons) exist together in an environment and improve each other's existence in some way. This fits holarchy-relationships very well. Any time one level starts to take advantage of other levels (a dysfunctional symbiotic action), eventually there will be a change in the holarchy that restores balance and symbiotic relationships. Otherwise, the holarchy will collapse, at least at the level of dysfunction and the later stages. For example, if an animal eats a diet that damages living cells in its body, deterioration of the animal's body will occur, leading to ill health or death.

The human race has become a dysfunctional level in our holarchy of life by raping our biosphere with much disregard of the results. If we don't soon learn from nature, our holarchy of life will collapse from underneath us and humans will no longer be a viable level in this biosphere-holarchy. It is like committing suicide out of ignorance. As the ignorance is beginning to end (as we understand how the chain of life works) we have the chance still to correct our actions and embrace our biological holarchy as part of whom and what we are.

When evolution is occurring (always) there is an order to the overall motion of things. That is built in. You might think of it as a probability. At any given moment, the action of evolution is unpredictable, but there are probable outcomes. This keeps uncertainty a part of the equation and that is an important factor. Otherwise it would just be like a mathematical function, or, in other words, a destiny. Destiny, where the future would be determined, is not the way nature works.

So, with humankind's current evolutionary step (a quantum leap size step), success is not guaranteed, but it is probable. We can still muck it up. If we do, it is highly possible that humans will survive, but it may not be a civilization that survives, just the species. It could put things back several thousand years and the survivors would start all over again. In the time frame of the universe, it is like a second. In our history, it would be the end of history with eventually a new start and a new history. It could happen more than once, until we get it right.

It is also possible, not probable, that humankind would become extinct. In that case, it would put things back hundreds of thousands or even millions of years. A new species of sentient beings would eventually arise and face this same level we are facing now. The solar system is young enough for this to happen a number of times.

We all have our own sense of reality and it's normal to think of ours as right and suspect others when theirs is different. But we all have a reality based on our current level of cultural development and it is limited by where we have come from in life. No one has the ultimate reality fully in place. Being able to examine our reality, and looking at other realities with an open mind and a willingness to grow in our understanding, helps us move forward with change and human evolution. It also helps to agree to grow by keeping an open mind and being flexible. Human culture has grown from the times of superstition and magic through the times of black and white traditional and is still moving toward greater broader realities. Joining that trend is moving with the nature of evolution rather than resisting it by holding on to the past. It is much easier to go with the river than struggle against the flow. We still end up down river either way.

As we understand the trends of evolution, we now have the opportunity to work toward the goals indicated by those trends, believe in that future and work to make it be.

Continuing Change

Lately there has been interesting developments in **neuroplasticity**. Neurobiologists are excitedly discovering the truly amazing adaptability of the gray matter in our heads. And what they are excited about is the realization that we can change – that grown adults have the capacity to develop and grow in many ways, even in our elder years. This is indeed very good news[1].

But the fact is, in spite of our neurological capacity to develop; many adults rarely do change in significant ways. More often than not, our adult years are about unending repetition rather than fresh learning and growth. It is possible for adults to continue personal cultural development throughout their lives. Living longer lives in recent generations have provided opportunities for substantial further development in later years.

Maslow's stages of needs show areas where cultural development can occur in adults. In our current society, it is common to think of human development as something that happens to children growing up. It is, however, something that adults can continue doing throughout the rest of their lives. Some people indeed do, but it is not yet the norm. Once people enter the workforce, they all too often get caught up in the struggle of making a living and don't think about further developing. People who continue learning and growing generally are happier with life. One of the tendencies of nature is if you are not moving forward, you tend to drift backward, and that is not fun.

Sir Isaac Newton's third law of motion says that for every action there is an equal and opposite reaction. This applies to all physical objects. In a way, it also applies to our mind. Our experience of life is determined by our minds evaluation of events. How our mind does that is through our subconscious. As young children, our subconscious is programed from our experience. Later, as adults, that programing creates our reactions to life and therefore our experience of it. An example of this is when a new president is elected, some people are happy about it while others think it is a disaster. It's the same event but with very different reactions. One person experiences joy, another sadness. The reason is the different personal world views, which are a result of each person's subconscious programming. The point is: What we put into our subconscious is what determines our experience of life. We program our subconscious through our day to day conscious thinking. A story from an unknown source goes: a native American grandfather talking to his grandchild says "there are two wolfs inside me fighting. One is the wolf of love and peace the other is the wolf of anger and war." The grandchild asks "which one will win." The grandfathers answer, "The one I feed."

This is where the equivalent of Newton's third law comes in, as the saying goes, what goes around comes around. Regular negative thinking programs us for negative experiences from life and that's what results. It's garbage in, garbage out. It can also be quality in, quality out. Hateful people experience life as hateful and attract hateful people into their lives. Positive people have more positive experience of life, are happier and even tend to live longer.

Once people realize that brains and personalities are not irrevocably fixed by the time we reach adulthood, we can actually improve. By reducing intake of negativity (from sources like TV, news and negative people) mental environments can improve. By watching our thinking and stopping negative ideas when they show up, we start to improve our subconscious programming. In this way, we have the opportunity to grow into a better future rather than just complaining and suffering about the changes.

When all is said and done, it is the evolutionary path that remains. This path is the one steady force behind all things. It is the impulse that explains life, the universe, human history and the future that humans are evolving into. It is the theme of material existence. All the politics, problems, sciences, religions and the daily struggles of humanity are, in the long view, relatively unimportant when seen against the backdrop of evolution. The daily struggles we experience are like one typical moment in the life of a person. At the end of that life, how significant was that one moment of struggle? Yet during that moment, nothing else may have seemed important. This helps put the current moment's struggle into perspective.

As we understand the trends of evolution, we now have the opportunity to work toward the goals indicated by those trends, and work to make it be.

World Peace?

World peace has been a dream as far back as known. Is it a hopeless dream? It can't occur as long as significant portions of world population live in abject poverty and are without hope. That is a recipe for terrorism and revolutions. Only when the global thinking is dominated by caring for others and making sure no one is left behind, that all have a reasonable semblance of what is needed to live a decent life, can real world peace happen.

The trends in evolution looked at in this book show that not only is world peace possible but probable. In fact, it is the only long term possibility for the human race. Anything less leads eventually to ruin and extinction. We can only ignore the trends of evolution for so long before they catch up with us. Now that we are learning about the big picture of evolution, we are getting the tools to think and act smarter. And this is just in time.

To have any chance of world peace, we need a true global village, which involves a higher order holon, a higher level of complexity. To sustain this higher order of complexity requires a greater level of interconnectedness, communication and feedback loops. With the internet came the interconnectivity. With cell phones, and especially smart phones, the communication and feedback loops are now spreading everywhere. The new level of holon can now include and transcend the old while producing new emergent qualities. This too is just in time.

Conclusion

We are entering the point where, as a species, we are aware of the need for periodic evolutionary steps to avoid the slow process of stagnation and decay. Knowing that and understanding the long view of evolution, we have the opportunity to deliberately move forward before the stress levels become extreme and get too uncomfortable and dangerous. By continuous small steps we can reduce the need for big leaps. By deliberately taking the small steps we may avoid the chaos level of change.

It is appropriate here to repeat something from the preface. Several ideas in this book are not universally accepted by all of science, but then almost nothing in science is. Science is a constantly changing field of knowledge and ideas. What is true today is tomorrow's folly. It has happened in science that truth became folly and later "truth" again. It has also happened many times that one day's folly later became truth. I leave it up to you to decide which this book is. If you decide it is folly, read it again in a year or two. Things are changing that fast.

New ways of understanding invariably run into resistance by many who learned older ideas which they still believe. That does not make new ideas necessarily right, but it often makes them controversial. Several examples among many are plate tectonics (originally called continental drift), the Big Bang theory, and quantum mechanics. Each was soundly criticized and ridiculed in its infancy, yet they are a mainstay of today's science.

At the incredible speed of today's progress, experts in their field often are experts in what was. It takes an open mind to look beyond old ideas and be able to see what may be. This book is an attempt to do that. I am quite confident that some of what I have written will end up being wrong. I expect that most of it will be shown to be accurate in the long run, but it may take 30 years to be sure.

If you liked some of the ideas presented in this book but are still having a hard time accepting the transformation of humanity, or if you would like to accept it just because it sounds good, I suggest you wait, watch and evaluate. After a few weeks, re-read this book starting at chapter 4. After several months, reread it again starting at chapter 7. That will put the ideas into long-term memory, where you can easily access them later. As events move along, notice when something new seems to crop up. If it is part of the transformation, and you notice it, you will then know. Then re-read the book again, starting at chapter 4. You will then be ahead of most people in understanding the forces at hand. May you live in interesting times.

Epilogue

You have brains in your head. You have feet in your shoes. You can steer yourself any direction you choose. – Dr. Seuss

It is the growing dysfunction of our current world order that has pushed us past a bifurcation point and into chaos. We have moved beyond homeostasis and into decay. The threat of our species' demise is real. To move into a new tier of human evolution takes a high level of stress. For the first time in human evolution, the environmental stress is a global threat and demands a global response. Only this can push us to a revolutionary evolutionary "quantum leap".

We are right where we need to be. It is the worst of times, and it is the best of times. We can cry in fear or laugh in delight. We can be sure we are going to hell or see humanity at the gate of heaven. It's our choice now, scream or rejoice. Those who scream the loudest will have the most pain in the transition. Those who rejoice will go much easier. Whether we succeed at the reorganization to a higher order or fumble into extinction cannot be foretold. The tendency of evolution suggests success but does not guarantee it.

At the present time, with over seven-billion people on earth and enormous change at hand, it may seem illogical to say the future of the human race is in your hands. However, the process of change (which includes bifurcation points and chaos), shows that at the critical point in time, very minor things (a butterfly flapping its wings in Brazil) can lead to huge effects (a tornado in Texas).

In other words, the effects of single human action, or small groups of people, can indeed have far-reaching effects at times like these. As Margaret Mead has said, "Never doubt that a small group of dedicated people can change the world, in fact it is the only thing that ever has". From a trickle of water came a trend, from the trend came the Grand Canyon. Theodore Roszak said, "All revolutionary changes are unthinkable until they happen – and then they are understood to be inevitable." We are now at an inevitable point in world cultural development.

Not to worry, it's just a little avalanche way up there. We can sit and watch what happens or we can join in the chaos and be part of what happens. If you could imagine a grandchild asking you someday, "what did you do when the human race was in danger?" what would you say? It is a time to do more than just watch. There is too much at stake. It is time to push back against business as usual. The course of what is to come will be determined by what people do now, not just what they think about, but what they do. It is a time for doing like none before. The opportunity will be short lived. Will we?

The evolution of the cosmos has been one of ever-increasing complexity, of ever increasing cooperation and communication of its parts, and growing integration. We are at the tip of almost fourteen billion years of evolution and are evolving for the first time ever with the self-knowledge of this fact. Evolution is not just a subject to think about, it's going on right now in the streets and it can be experienced in the raw, in motion. No human has ever experienced this before. In the past, new stages happened so slowly that by the time they were noticed as such, those who were alive at the start were long gone before the middle. We're in new territory.

The fact that things are changing faster than in the past is obvious to almost everyone. It is easy to believe that the trend will continue. It is much more difficult to appreciate the near future of explosive exponential change. The approaching chaotic change will make all normal events minor in comparison. On the other hand, such drastic change is exactly what it takes to move humanity through its crisis of adolescence and into adulthood. That is what is needed now to solve the global mess we are in. Holding on tightly to the past will be painful. Living the change like a teenager on a roller coaster could even be fun.

We are just now beginning to really understand the big picture of evolution and how it works. Knowing this and knowing that we are living in its midst, we have the unique option of planning our evolving future. Instead of evolving by accident or happenstance, we can do it by intelligent planning thus avoiding some of the unpleasant missteps. Thinking about the significance of that is mind boggling.

One of the significant things about understanding these ideas is as follows: Emerging science and technology may provide the tools to solve current problems, but by itself will not change the level of thinking that created the problems. If a group of people worked hard for many years to solve a major world problem, like say global warming, and succeeded, the other major problems would remain and have gotten worse. If a group instead worked hard for many years to advance the level of cultural development and succeed, they solve all the major global problems at one time without specifically working on any one. This is because they would change the level of thinking to a higher order than the level that created the problems. The current level of cultural development has reached the point of stagnation and pathological problems are growing. Only by reorganizing into a higher level of complexity and its emergent qualities can we successfully use the tools of science and technology to move beyond today's global crisis.

The human race is now beginning to awaken to its potential, just beginning to understand what can be. Moving beyond being driven by needs, it is poised on the threshold of maturity. The potential future can be grand beyond words. Our past is not our future. We are at a point of departure, where the trends of the past can only tell us that our future will be something notably different. Listen gently to the echoes of the future.

Let us ask the question: "What is it that we want to become?" Let us begin. NOW Is The Time.

Glossary

ABB: Used to designate a time span After the Big Bang.

AQAL: AQAL is an integral theory acronym for All Quadrants, All Levels. It also implies all types, stages and lines that apply. It simply means having considered and integrated all aspects of the whole picture using the full integral theory spectrum.

CE: Common Era, the years after BCE, the same meaning as AC. CE and BCE are used in science rather than AC and BC, which have religious implications and are of only western culture, not eastern.

Adaptive Systems Theory: Is the theory about groups of interacting or interdependent items that form an integrated whole and contain feedback loops. The system is able to respond to changes in its environment. The feedback loops provide the communications that allow the system to adapt.

Anthropomorphic: Human like, or ascribing human like qualities to nonhuman things. An example is assuming God has qualities like humans, such as being temperamental or having a gender.

BCE: Before the Common Era, before the CE years, the same meaning as BC. CE and BCE are used in science rather than AC and BC, which have religious implications and are of only western culture, not eastern.

BI: Before the Internet, which was open to the public in 1989.

Bifurcation point: Bifurcation means to split into two branches, like a fork in the road. Used in this book, it is more specific. A bifurcation point in dissipative structures is a point where environmental stress has pushed a self-adaptive system beyond its capability to adjust, producing a time of transition.

Billion: This book uses the US definition which is 1,000,000,000.

Break Point: The point after a bifurcation point has been reached and stress levels have continued to rise where the onset of chaos occurs.

Cartesian Mechanism: Historically Cartesian mechanism is a philosophy, developed in the 1600's, which explains physical and biological phenomena as only aspects of mechanical effects of matter. No religious or metaphysical aspects are needed. It is the basis of scientific materialism.

CEO: Chief Executive Officer, the top executive of a corporation.

Chaos Theory: A scientific theory where very small changes in initial condition can produce very large and unpredictable changes later on in the event, such as weather predictions from computer models.

Cosmology: The study of the universe (cosmos) in its totality.

Culture: As used in this book, culture refers to the overall set of patterns that a group of people live by, their views and values, the attitudes and behaviors that are their common characteristics.

Glossary

Cultural Creatives: A cultural group (also called "postmodern") first appearing in the 1960's, and characterized by being sensitive to the environment, equal rights, consensus decision making and being self-improvement oriented, among other traits.

Depth: Used in integral theory, it is the distinction of which level on a line that a holon is on. It is measured as increasing as it moves away from the center of the quadrant. A molecule is at greater depth than an atom, but less depth than an organism.

Determinism: The term for a belief of science that when the details are known about something at one point in time, that it predictably determines what will happen in the future and what happened in the past. Sir Isaac Newton's laws of motion fortified the idea of determinism in the 1600's.

Dissipative Structures Theory: A theory that applies to all living systems as well as many non-living systems. It describes how systems maintain order until a point is reached when they evolve into something more or decay and dissolve.

Dogma: The system of principles, laws, practices or rules established by an organization, such as a religion.

Dualism: The philosophical idea that there are two basic elemental substances of reality, material and mental.

Elephant: A large mammal that has at times caused considerable confusion among groups of blind people.

Emergent: Emergent refers to qualities of a thing that didn't exist in its separate parts but emerge, or come into existence when the parts combine into a new thing. A car has qualities, such as drivability and handling that its parts don't separately have.

Entropy: A thermodynamic term that refers to the loss of energy or order of a system over time due to things like friction or loss of heat.

Eukaryotes: Living cells that contain a nucleus that contains DNA.

Exponential: The name of a curve of data that has a regular growth rate. For example, if savings grows at 5% each year the cumulated savings is exponential.

Facebook: An internet site for communicating information and photos with networks of friends.

Global Acres: A measure of sustainability that uses an average worldwide acre of land to produce an average amount of food, natural materials, housing, and infrastructure (such as fresh water, waste disposal etc.) to support the life style of an average human being.

Global Climate Change: is the change in climate patterns due to the increase in the average temperature of the Earth's air and oceans since the mid-20th century. Most of the observed temperature increase since the middle of the 20th century has been caused by increasing concentrations of human generated greenhouse gases from fuel burning and deforestation.

Greenhouse Gases: The main greenhouse gases are carbon di-oxide, ozone, methane, nitrous oxide and CFCs. These cause the greenhouse effect which is the process by which the absorption of solar radiation by the gases warms the lower atmosphere and surface.

Glossary

Holoarchy: A Holoarchy is a natural hierarchy. A holoarchy is a hierarchy of holons and is free of judgments about one level being better than another as is often found in cultural hierarchies.

Holon: A Holon is a whole thing in and of itself while also being part of a larger whole and being made up of smaller whole things. A molecule is an example. It is part of a cell, a larger whole, and is made up of atoms, smaller whole things.

Homo: The genus homo includes modern humans as well as several extinct related species such as Homo erectus and the Neanderthals.

Homeostasis: Maintaining a relatively stable state of equilibrium within a system.

Inflection Point: On a plot of data, the point on the curve where the curvature changes from convex to concave or vice versa.

Integral Theory: Integral theory takes all cultures and science disciplines into account in producing an integrated way of organizing things into a single unified framework. "It is, at once, ancient and modern, Eastern and Western, speculative and scientific, and yet also something beyond those dichotomies. integral means 'comprehensive, whole, and balanced.' It's a synthesis of the 'best of the best' that our traditions have to offer, combined with the most state-of-the-art transformational techniques"[1]. It explains how many things that seem to be opposites are actually parts of the whole.

Level: In integral theory, a level, also called a stage, represents a condition of growth or development. For example, levels within a grade school are grades 1 through 6.

Line: In integral theory, a line represents a holarchy with its start at the center of the four quadrants and moving away within one of the quadrants. A line will have levels or stages along its length.

Materialism: The theory that matter is the only reality, that all things are material including thought, mind and feelings.

Medium Change: Change that occurs after the environmental stress has reached a bifurcation point but has not risen to the chaos level. It is more than small change and less than transformational change.

MEME: A term for the stages of development used in spiral dynamics theory. A MEME reflects a structure of world view and values, and as such the way people think.

Mitosis: The process where a living cell reproduces by dividing into two cells, each of which has the same DNA as the original.

Modern Science: A term referring to the scientific method and set of beliefs that arose in the 1700's.

Modernism: Of modern times and character, usually 20th or 21st century and sometimes the 19th century.

Modernity: The western cultural world view starting in 1543 CE and continuing up to recent times.

Nanotechnology: The prefix "nano" refers to extremely small, a billionth. Nanotechnology generally means a product made using elements (such as atoms, molecules or electrical components) in the general order of a one to 100 billionth of a meter.

Glossary

Normal Change: The kind of day-in and day-out change we experience in normal life all the time, such as day changing into night and the constant change of the clock.

Open system: An open system is one that interacts with its environment as opposed to a closed system that does not.

Paradigm: Something that serves as a pattern or model, a set of assumptions, concepts, values, or practices that is a way of viewing reality. Example, Sir Isaac Newton's theories in physics changed the paradigm of his time to a new view on how the world worked.

Prokaryotes: Living single cell organism representing the most primitive form of life. Prokaryotes do not contain a nucleus.

Quadrant: One of four parts of a plane, separated by two perpendicular lines. In integral theory, the four quadrants represent the inner-individual, inner-collective, the outer-individual and the outer-collective. Together, everything fits into one or more quadrants.

Reductionism: Is a scientific belief, originating in the 1600's, that you can understand complex things by reducing them to the interaction of their parts.

Renaissance: As used in this book, a cultural movement that profoundly affected European intellectual life in the early modern period. Beginning in Italy, and spreading to the rest of Europe by the 16th century, its influence affected literature, philosophy, art, politics, science, religion, and other aspects of intellectual inquiry. It spanned roughly the 14th to the 17th century. One development of the era was the scientific method. This revolutionary new way of learning about the world focused on empirical evidence, the importance of mathematics, and establishing a mechanical philosophy of nature. The term "renaissance" actually has many meanings and defies simple explanation.

S-curve: An S-curve is a plot of data that has a curve shaped something like the letter "S". It is also called a Logistic Curve in some uses.

Sapiens: Used in Homo sapiens means knowing or self-knowing humans.

Self-Organizing System: An open system that has a network of feedback loops such that it can change its actions to maintain its structure without external guidance.

Singularity: In cosmology, a singularity is a point where the laws of physics do not work, where the mathematics will not give rational results. Such a place was the original point where the Big Bang started and also around black holes.

Small Change: Change brought on by a small stress level that leads to a system easily adapting in such a way as to be able to handle the stress level normally.

Stage: In integral theory, a stage, also called a level, represents a condition of growth or development. For example, stages within a grade school are grades 1 through 6.

State: In integral theory, a State is a condition of a thing with respect to circumstances or attributes. Examples include a state of mind, state of health or state of matter.

Strange Attractor: The name given for an area on a plot of chaotic data around where the data are oriented.

Supernova: A massive star that undergoes a sudden massive explosion at the end of its life cycle as a result of gravitational collapse. The star ends up as a black hole or super dense neutron star.

Sustainability: In a broad sense, sustainability is the capacity to endure. In ecology, the word describes how biological systems remain diverse and productive over time. For humans it is the potential for long-term maintenance of wellbeing, which in turn depends on the wellbeing of the natural world and the responsible use of natural resources.

Symbiosis: Two or more organisms living together in a close relationship where each provides something of value to the other. An example is in the human body, bacteria thrive on food provided by the host (a person) while the host needs the bacteria to digest its food. Without the bacteria, a human would starve. This use of the word symbiosis is also called "mutualism".

Transformational Change: Major change driven by high stress levels that leads to a substantial and significant change in the system in order for it to be able to function normally at the higher stress level.

Transition: The process of moving from one system configuration into another, driven by external stress. This is a three step process that includes the end of the old configuration, an in-between period, and the birth of the new configuration.

Type: In integral theory, a Type is a characteristic description of a thing. Examples include male or female as types of gender; and train, bus or plane as types of transportation.

NOTES

Chapter 1

[1] Albert Einstein, Quote DB, http://www.quotedb.com/quotes/11.

[2] Thomas Summer, "Global warming continues apace" *Science News*, December 13, 2015, 21.

[3] The Daily Censored Underreported news and Commentary, http://dailycensored.com/2009/11/02/is-global-warming-a-corporate-fraud/.

[4] Global Issues, http://www.globalissues.org/article/710/global-warming-spin-and-media.

[5] National Geographic, *EarthPulse State of the Earth 2010*.

[6] Harvard Gazette, *Harvard Finds Kidney Stones, Malaria Among Global-Warming Risks,* November 2009, http://news.harvard.edu/gazette/story/2009/11/harvard-finds-kidney-stones-malaria-among-global-warming-risks.

[7] Sciencestage.com, *NASA data: Greenland, Antarctic ice melt worsening*, http://sciencestage.com/resources/nasa-data-greenland-antarctic-ice-melt-worsening-ap-0.

[8] The Observer, *Scientists to issue stark warning over dramatic new sea level figures*, March 8, 2009, http://www.guardian.co.uk/science/2009/mar/08/climate-change-flooding.

[9] Planet Ark, *Sea-Level Rise Poses New Flood Risk To California*, Mar 12, 2009, http://planetark.org/ark/51995.

[10] American Heart Association, *Air Pollution, Heart Disease and Stroke*, http://www.americanheart.org/presenter.jhtml?identifier=4419.

[11] Theworldcounts, *Hazardous waste statistics*, http://www.theworldcounts.com/.

[12] Rachel Ehrenberg, *Styrofoam degrades in seawater*, Science News, September 12, 2009; Vol.176 #6, page 9.

[13] Charles Q. Choi, "Ocean Acidification from CO_2 Is Happening Faster Than Thought", *Scientific American*, February 2009, http://www.scientificamerican.com/article.cfm?id=ocean-acidification.

[14] Charles Duhigg, *Toxic Waters*, The New York Times, October 15, 2009, http://projects.nytimes.com/toxic-waters.

[15] Jane Braxton Little, "Welcome to the Sixth Mass Extinction", Discover Magazine, January/February 2016, 24.

[16] Ellen W. Demerath, *Human Biology*, Project Muse, http://muse.jhu.edu/login?uri=/journals/human_biology/v075/75.1demerath.html.

[17] Chris Hails, ed., *Living Planet Report 2008*, World Wide Fund for Nature (2008). http://assets.panda.org/downloads/living_planet_report_2008.pdf.

[18] Thom Hartmann, *Threshold* (New York: Penguin Group, 2009), 24-25.

[19] The Economist, *The World in 2009*, 112.

[20] Thomas Sumner, "Many of Earth's groundwater basins run deficits", Science News, July 25, 2015, 13.

[21] Alan Robock and Owen Brian Toon, "Local Nuclear War - Global Suffering*", Scientific American*, January 2010, 74-81.

[22] Source: http://www.epa.gov/climatechange/ghgemissions/global.html

[23] National Oceanic & Atmospheric Administration, *Mauna Loa Observatory*, http://www.esrl.noaa.gov/gmd/obop/mlo/.

Chapter 2

[1] National Geographic *EarthPulse State of the Earth 2010, 34.*

[2] US Environmental Protection Agency, http://www.epa.gov/waste/nonhaz/municipal/pubs/msw2008rpt.pdf.

[3] *The Agonist* Thoughtful, Global, Timely. http://agonist.org/the_5_fundamental_problems_of_the_us_economy.

[4] Thom Hartmann, *Threshold* (New York: Penguin Group, 2009), xiii.

Chapter 3

[1] Jonathan Barnes, *Early Greek Philosophy* (New York: Penguin Group, 2001), 7-17.

[2] Encyclopaedia Britannica, *Black Death*, http://www.britannica.com/event/Black-Death.

[3] Edward Grant, *The Foundations of Modern Science in the Middle Ages*, Cambridge 1996, 27.

[4] Edmund J. Bourne, *Global Shift* (New Harbinger Publications, Oakland, 2008), 36.

[5] Francis Bacon, Novum Organum, trans. and ed. P. Urbach and J. Gibson (Peru, IL: Open Court, 1994).

[6] Encyclopaedia Britannica, *Johannes Kepler*, http://www.britannica.com/biography/Johannes-Kepler.

[7] Edmund J. Bourne, *Global Shift*, 37.

[8] Ibid., 38.

[9] Baptist Planet, *Religious Connections*, http://www.baptistplanet.com/2009/10/massive-study-of-global-religious.html.

[10] Rapid City Journal, *Trends in Church Attendance Spell Trouble for Faiths*, http://www.rapidcityjournal.com/lifestyles/faith-and-values/religion/article_ebd80dcc-6b8d-11df-af11-001cc4c002e0.html.

[11] John Mark Ministries, *Does The Australian Church Have a Future?*, http://jmm.aaa.net.au/articles/8517.htm.

[12] Times on Line, *Churchgoing on its knees as Christianity falls out of favour*, http://www.timesonline.co.uk/tol/comment/faith/article3890080.ece.

[13] Pablo Branas-Garza, *Church attendance in Spain (1930-1992)*, http://ideas.repec.org/a/ebl/ecbull/v26y2004i1p1-9.html.

Notes

[14] Christian News Watch, *How Many People go to Church in the UK*, http://www.whychurch.org.uk/trends.php.

[15] David Ewart, *United Church of Canada People Trends*, http://www.davidewart.ca/2009/02/united-church-of-canada-people-trends.html.

Chapter 4

[1] Encyclopedia Britannica, *The 20th-century revolution*, http://www.britannica.com/EBchecked/topic/528771/history-of-science/29341/The-20th-century-revolution.

[2] Facebook, http://www.facebook.com/.

[3] Ludwig von Bertalanffy, *General System Theory* (New York: George Braziller, Inc., 1969), 90.

[4] Fritjof Capra, *The Web of Life*, (New York: Anchor Books, 1996), 29.

[5] Ibid., 222-244.

[6] Edward Lorenz, "Deterministic nonperiodic flow". *Journal of Atmospheric Sciences*. 20, (1963): 130-141.

[7] Massachusetts Institute of Techology, *Edward Lorenz*, http://news.mit.edu/

[8] Capra, *The Web of Life*, 82.

[9] Walter Buckley, *Society – A Complex Adaptive System* (Amsterdam: Overseas Publishers Association,1998), 99.

[10] Ilya Prigogine, *The End of Certainty* (New York, The Free Press, 1996), 66.

[11] Ibid., 66-72.

[12] Capra, *The Web of Life*, 191.

[13] Robert Frager and James Fadiman, *Personality and Personal Growth* (Upper Saddle River, New Jersey, Pearson Education, 2005), 345-47.

[14] Ibid., 334-348.

[15] Steve McIntosh, *Integral Consciousness* (St. Paul, MN, Paragon House, 2007), 31.

[16] Don Edward Beck and Christopher C. Cowan, Spiral Dynamics (Malden, MA, Blackwell Publishing, 1996), 31-68.

[17] McIntosh, *Integral Consciousness*, 34-56.

[18] Paul H. Ray and Sherry Ruth Anderson, *The Cultural Creatives* (New York, Three Rivers Press, 2000), 1-40.

[19] Paul H. Ray, *The Potential for a New, Emerging Culture in the U.S.*, https://www.wisdomuniversity.org/CCsReport2008SurveyV3.pdf.

[20] Frank Wilczek, *The Lightness of Being* (New York, Basic Books, 2008), 30.

[21] Ibid., 131-2.

[22] Ibid., 104.

[23] Ibid., 90-92.

[24] Jeremy L. O'Brien, *Exploiting Entanglement* (Science News, October 29, 2010), 588.

[25] Ron Cowen, "Further Evidence for Dark Energy," *Science News*, April 9, 2011, 16.

Chapter 5

[1] Sidney Liebes, Elisabet Sahtouris and Brian Swimme, *A Walk Through Time* (New York, John Wiley & Sons, Inc., 1998), 70.

[2] William Bridges, *Transitions* (New York, Addison-Wesley Publishing Co, 1980), 18.

Chapter 6

[1] Bridges, *Transitions*, 8-18.

Chapter 7

[1] Ken Wilber, *Integral Vision* (Boston, Shambhala Publications, 2007).

[2] Ken Wilber, *A Brief History of Everything* (Boston, Shambhala Publications, 1996).

[3] Arthur Koestler, *The Ghost in the Machine* (reprint edition, Penguin Group, NY, 1990).

[4] http://www.holacracy.org

Chapter 8

[1] Fritjof Capra, *The Web of Life* (New York, Anchor Books, 1996), 222-225.

[2] Sidney Liebes, Elisabet Sahtouris and Brian Swimme, *A Walk Through Time* (New York, John Wiley & Sons, Inc., 1998), 110-112.

[3] Gould, Stephen J, *Ever Since Darwin: Reflections in Natural History* (New York, Penguin, 1977).

[4] Gould, Stephen J, *The Mismeasure of Man* (New York, W.W. Norton & Company, 1996).

[5] Tina Hesman Saey, "Scientist Still Making Entries in Human Genetic Encyclopedia", *Science news*, November 6, 2010.

[6] Mihaela Pertea and Steven L Salzberg, "Between a chicken and a grape: estimating the number of human genes", Genome Biology, 2019, 11 (5 May 2010): 206.

[7] Bruce H. Lipton and Steve Bhaerman, *Spontaneous Evolution* (New York, Hay House, 2009), 130-136.

[8] Ibid., 148-156.

[9] Martin Novak and Roger Highfield, *Super Cooperators* (Free Press, NY, 2011), 268-281.

[10] Ibid, 280.

[11] E. Toby Kiers and Stuart A. West, "Evolving new organisms via symbiosis" *Science 348* (April 24, 2008), 392-394.

[12] Frank Wilczek, *The Lightness of Being* (New York, Basic Books, 2008), 32-33.

[13] Sidney Liebes, Elisabet Sahtouris and Brian Swimme, *A Walk Through Time* (New York, John Wiley & Sons, Inc., 1998), 14.

[14] Ron Cowen, *Windows on the Universe* (Science News, October 10, 2009), 17-20.

[15] Peter Russell, *Waking Up in Time* (Novato, CA, Origin Press, 1998), 167-168.

[16] Liebes, *Walk Through Time*, 26.

[17] David Christian, *Maps of Time* (Berkeley, CA, University of California Press, 2011), 79.

[18] Liebes, *Walk Through Time*, 40.

[19] Ibid., 34.

[20] Ibid., 36-53.

[21] Ibid., 70-80.

[22] Capra, *The Web of Life*, 230-231.

[23] Liebes, *Walk Through Time*, 89-92.

[24] Russell, *Waking Up in Time*, 13.

[25] Ibid., 14.

[26] Liebes, *Walk Through Time*, 120-132.

[27] Ibid., 272.

[28] Capra, *The Web of Life*, 258-259.

[29] Ann Gibbons, "A New View Of the Birth of Homo sapiens," *Science* 331, (28 January 2011): 394.

[30] Russell, *Waking Up in Time*, 15.

[31] Ibid., 18.

[32] Ron Cowen, "Further evidence for dark energy," *Science News*, April 9, 2011, 16.

[33] Tina Hesman Saey, "Human body not overrun by bacteria," *Science News*, February 6, 2016, 6.

Chapter 9

[1] Robert Kurzban and H. Clark Barrett, "Origins of Cumulative Culture," *Science* 335, (2 March 2012): 1056-57.

[2] Kin Wilber, *Integral Psychology* (Boston, Shambhala, 2000), 39-44.

[3] Don Edward Beck, Christopher Cowan, *Spiral Dynamics* (Malden, MA, Blackwell Publishing, 1996), 197-202.

[4] Wilber, *A Brief History of Everything*, 68.

[5] Beck, *Spiral Dynamics,* 203-214.

[6] Ibid., 215-225.

[7] Ibid., 229-237.

[8] Steve McIntosh, *Integral Consciousness* (St. Paul, MN, Paragon House, 2007), 44.

[9] Ibid., 23-30.

[10] A. C., "The Power of True Believers", *Science* 332, 8 April (2011): 171.

[11] Don Edward Beck and Christopher C. Cowan, Spiral Dynamics (Malden, MA, Blackwell Publishing, 1996), 65.

[12] McIntosh, *Integral Consciousness,* 153-193.

[13] Ken Wilber, *Creating An Integral Culture*, http://www.beyondawakeningseries.com/blog/archive/, 16 September 2010.

[14] Rafe Sagarin, "Sink or Swim," *Wired*, (April 2012): 21-22.

¹⁵ Peter H. Diamandis and Steven Kotler, *Abundance, the Future is Better Than You Think*, (Free Press, NY, 2012).

¹⁶ Matt Ridley, *The Rational Optimist: How Prosperity Evolves* (Harper-Collins Publishers, NY, 2010).

Chapter 10

¹ Leslie Roberts, "9 Billion?," *Science* 333, (29 July 2011): 540.

² Russell, *Waking Up in Time*, 24.

³ Ray Kurzweil, *The Singularity Is Near* (the Penguin Group, NY, 2005), 63.

⁴ Brian Vastag, The Washington Post, *Exabytes: Documenting the 'digital age' and huge growth in computing capacity*, http://www.washingtonpost.com/wp-dyn/content/article/2011/02/10/AR2011021004916.html

⁵ Russell, Waking Up in Time, 18-19.

⁶ Larry Roberts, *Internet Traffic Growth*, http://www.netvalley.com/intvalstat.html

⁷ David Christian, *Maps of Time* (University of California Press, California, 2005), 79-82.

⁸ Ray Kurzweil, *The Singularity Is Near*.

⁹ David Christian, *Maps of Time*.

Chapter 11

[1] Dmitry Kucharavy and Roland De Guio, "Logistic Substitution Model and Technological Forecasting", *Triz-Journal*, (February 2, 2009).

[2] Theodore Modis, "Strengths and weaknesses of S-curves," *Technological Forecasting and Social Change*, Vol. 74, No. 6, (July 2007), 866-872.

[3] Sidney Liebes, *A Walk Through Time*, 48, 56.

[4] Ibid., 36.

[5] Ibid., 68-74.

[6] Theodore Modis, "Fractal Aspects of Natural Growth," *Technological Forecasting and Social Change*, Vol. 47, No. 1, (1994) 63-73.

[7] United Nations, *World Population to 2300*, (United Nations, NY, 2004), 5.

[8] Hod Lipson and Melba Kurman, *Fabricated*, 34-39.

[9] Diamandis and Steven Kotler, *Abundance*, 199-200.

Chapter 12

[1] Diamandis and Steven Kotler, *Abundance*, 105.

[2] Ibid., 106.

[3] Sky-High Vegetables: Vertical Farming Sprouts In Singapore, NPR, http://www.npr.org/blogs/thesalt/2012/11/06/164428031/sky-high-vegetables-vertical-farming-sprouts-in-singapore

[4] Diamandis and Steven Kotler, *Abundance,* 109.

[5] Kai Kupferschmidt, "Lab Burger Adds Sizzle to Bid for Research Funds", *Science*, 341, (2013): 602-03.

[6] Diamandis and Steven Kotler, *Abundance,* 174-177.

[7] Ibid., 184-186.

[8] Diamandis and Steven Kotler, *Abundance,* 195-196.

[9] Bill & Melinda Gates Foundation, *Global Health Division*, http://www.gatesfoundation.org/what-we-do

[10] Diamandis and Steven Kotler, *Abundance,* 85-97.

[11] Ibid., 167-173.

[12] Ibid., 279.

Chapter 13

[1] Laura Sanders, "Lifting neural constraints could turn back time, making way for youthful flexibility", *Science News*, (August 11, 2013), 19-21.

Glossary

[1] Ken Wilber, *Integral Life Practice* (Boston, Integral Books, 2008), 1-2.

Index

ABB, 98, 99, 100
Abraham Maslow, 56
accelerating, 7, 9, 15, 71, 127, 144
acidification, 10
adaptive systems, 49, 51
Adolph Hitler, 16
advertising, 5, 27
aerobic, 103
aeroponics, 158
air, 9, 18, 31, 102
airports, 8
Albert Einstein, 4, 19, 37
Animals, 104
Antarctica, 7
AQAL, 90, 91
archaic, 39, 58, 59, 112
Arthur Koestler, 83
atmosphere, 7, 10, 100, 102, 140, 141
atmospheric, 10
atoms, 56, 82, 83, 84, 98, 99, 108
bacteria, 51, 68, 75, 77, 95, 101-3, 108, 120, 122, 139, 140, 141
banned, 9, 43
bifurcation point, 54, 55, 66, 72, 75, 77, 95, 102
big bang, 61, 97
Big Bang, 97, 98, 100, 142, 143
big business, 8
Bill & Melinda Gates Foundation, 163
Bill Gates, 157, 166

billionaires, 118, 156, 162
biosphere, 140, 171
birth control, 29
birth rates, 13
birthrates, 168
black hole, 142
brain, 65, 105, 106, 111
break point, 67, 74, 77
breathers, 102
carbon dioxide, 10, 102, 103
caterpillar, 72, 74
CEO, 24-6
change, 9, 13, 16, 25, 29, 30-2, 47, 48, 54, 55, 57, 62-71, 73, 74, 93-5, 101, 107, 117, 119, 121-3, 127-9, 136, 139, 141-4, 147, 149, 150, 152
change of state, 142
chaos, 5-5, 67, 71-5, 77, 94, 95, 102, 108, 117, 121, 136, 149
chaos theory, 52, 53, 67, 108
chaotic, 53, 67, 71-4, 102, 108, 178
chaotic systems, 53
Charles Darwin, 50, 69, 94
chess, 134
church attendance, 40
civilization, 64
civilized, 18
Clare W. Graves, 57
classical, 3-5
climate, 10, 53, 75

cognitive, 56, 106, 111, 112
cold war, 16
collapse, 14, 15, 17, 64, 121
community, 28, 56, 58, 113, 115, 122, 144
compassion, 119
competition, 42, 96, 114
complexity, 50, 54, 56, 58, 60, 61, 67, 71, 72, 74, 101, 103, 105, 107, 108, 109, 111, 115, 117, 120, 121, 132, 143, 144
computer, 16, 23, 28, 47, 48, 50-3, 129, 131
consumption, 14, 18, 22, 23, 29
contamination, 10
cooperation, 32, 50, 96, 108, 114, 121, 148, 178
Copernicus, 35-7
corporate, 16, 17, 19, 24, 25, 27
cosmology, 61
critical mass, 117
cultural, 30, 32, 33, 39, 41, 42, 55, 57-60, 67, 72, 81, 84, 91, 108, 109, 111, 112, 114-6, 118, 119, 121-3, 128, 131, 134, 143, 144, 146, 149, 150, 152, 155, 172, 173, 178
cultural change, 32
Cultural Creatives, 59
cultural development, 55, 111, 112, 117
cultural stage, 57, 58, 152

culture, 13, 21, 24, 25, 30, 31, 32, 34, 39, 42, 48, 56, 59, 68, 76, 90, 106, 107, 114, 117, 119, 121, 145, 147, 152
deadly, 16
decay, 14, 51, 74, 75, 76
depleting, 13
desalinization, 164, 165
Descartes, 37, 38, 49
desperate, 16, 18
destabilizing, 15
determinism, 38, 95
developing countries, 14, 15
diapers, 10
dinosaurs, 105
disaster, 9, 13
discontinuities, 94
discontinuity, 142
disease, 7, 148, 163
diseases, 7, 9
disposable, 22, 23
dissipative structures, 54, 64-6, 73, 84, 107, 108
diversity, 50, 104, 114
DNA, 103
Don Beck, 57
dot-com, 156, 160
droughts, 6, 7
dualism, 38
economic, 17, 21-4, 26, 74
economy, 15, 21, 22, 26, 27
ecosystem, 10
ecosystems, 14
education, 148, 160, 162, 163, 167, 214
Edward Lorenz, 51

Index

emergent, 82, 85, 92, 103, 170
emerging, 38, 50, 59, 107
energy, 19, 51, 53, 60, 69, 97-100, 102, 103, 132, 151, 158, 159, 165, 166
energy flow, 97, 132
energy storage, 166
entropy, 51, 53
environment, 9, 14, 17, 23, 25, 54, 57, 66, 90, 95, 101, 148
environmental, 6, 7, 9, 11, 54, 64, 74, 75, 94, 95, 102, 122
Environmental, 9
erosion, 9
Eukaryotes, 103
evidence, 10, 13, 39, 94, 98, 146, 149
evolution, 50, 54, 56, 70, 74, 76, 77, 93, 94, 96, 97, 104-9, 111, 120-2, 134, 140, 143
executive, 25
expansion, 22, 113
exploit, 15
exploiting, 13, 14, 29
explosive, 122, 128, 131, 132, 136, 143, 144, 151, 156, 167, 178
exponential, 23, 134-6, 139, 140, 141, 143, 144, 146-8, 151, 155, 156, 178
exponential rate, 23
extinct, 171
feedback loop, 51, 54, 58, 64, 67, 84, 85, 92, 107

fish, 13, 14, 31
fishing, 10, 14, 30
flood, 6
floods, 6, 7
food, 10, 15, 16, 19, 30, 31, 64, 101-3, 108, 113, 141
food chain, 10
foot print, 23, 26
fossil, 14, 94, 166
fossil fuels, 13
fractal, 143
Francis Bacon, 36
Frank Wilczek, 60, 98
fresh, 7, 13, 14, 31
fundamentalist, 42
future, 4, 21, 104, 112, 120, 136, 137, 143, 144, 148-50, 155, 158, 159, 165, 169, 172, 174, 177-9
galaxies, 61, 75, 99, 108
galaxy, 61, 100
Galileo, 36-8
gene, 104
genes, 95
genetic, 94, 95, 104, 107, 108, 122
Georges Anderla, 131
glaciers, 7
global acres, 13
global climate, 7
global crises, 4
global warming, 7-10, 27, 114, 158
Global warming, 5, 7
greed, 16, 17, 25, 26
Greek, 33-5, 75, 193
greenhouse, 5-7

greenhouse gases, 5, 6
Greenland, 7
growth, 13, 14, 21, 23-5, 27, 32, 68, 71, 74, 75, 89, 121, 122, 127, 128-9, 132, 135, 136, 140, 141, 143, 151
habilis, 105
habit, 30, 31
hazardous, 9, 11
health care, 91, 114
hierarchies, 25, 39, 56, 84, 114
hierarchy of needs, 56
historical, 33
holarchy, 84, 85, 88, 89, 92, 108, 171
holistic, 59, 118, 146, 147, 148, 150
holons, 83-5, 88, 89, 108, 170
homeostasis, 64, 66, 73-7, 177
Homo, 105, 106, 112, 152
human era, 107
humans, 10, 30, 35, 58, 64, 75, 77, 104-8, 111, 112, 119, 122, 128, 140
hunger, 16, 158, 160
hunter-gather, 30, 112
hurricanes, 6
hydrogen, 82, 83, 98, 99-102, 108, 140, 141
hydroponics, 158
illiterate, 28
Ilya Prigogine, 53
India, 17, 32
industrial revolution, 114

inflection point, 140, 141, 144, 146, 147, 150
information, 48, 49, 107, 115, 131, 150, 161, 162, 167
infrastructure, 15, 144
inquisition, 36
instability, 16
integral stage, 59, 115, 118, 119, 122, 146
Integral Theory, 61, 62, 85, 87-92, 94
internet, 48, 132, 145, 150
inventions, 129
irrigation, 15
Isaac Newton, 37, 38, 50, 52, 61
Josef Stalin, 16
Ken Wilber, 81
Kepler, 37
Khan Academy, 162, 163
killed, 16, 34
knee, 134, 136, 156
knowledge, 26, 35, 131
language, 106, 111
Leibniz, 38, 50
levels, 83, 89, 90
life, 17, 68, 77, 99-104, 107-9, 119, 122, 139, 140, 171
life expectancy, 13
lily, 134
limitations, 11, 145
linear, 50, 53, 108, 148, 150, 155
linear equations, 50, 53
lines, 83, 88
literacy, 34
Lyme, 7

malaria, 7
mammals, 105, 108
management theory, 90
Mao Zedong, 16
map, 4, 82, 89, 119
materialism, 38
medieval, 33-5, 39, 41, 104
medium change, 65, 68, 73
Melinda Gates, 157
melting, 7, 143
mental, 111
metabolism, 101
metamorphosis, 72
methodology, 37
Miami, 8
middle class, 114, 167
Millennial Generation, 40
mind set, 49
mitosis, 104
models, 5, 7, 10, 53, 90, 112
modern, 29, 33-9, 41, 42, 105, 106, 112, 114
modernist stage, 58
modernity, 35, 41, 42
Moore's Law, 129
mortality, 13, 168
multicellular, 103, 104
Nanotechnology, 164
NASA, 7
nature, 9, 11, 23, 30, 38, 48, 50, 56, 63, 64, 68, 73, 74, 77, 83, 84, 93, 96, 99, 102, 108, 119, 139, 141, 143, 144
Neo-Darwinism, 94
network, 51

neuroplasticity, 172
neutron star, 142
non-linear, 50, 51, 53, 54, 108
nonreplicable, 13
normal change, 65
nuclear power, 166
nuclear weapons, 16
nuclear winter, 16
oceans, 9, 10, 14, 31, 100, 102
oil, 10, 14, 18
OLPC, 162
open system, 51
optimistic, 19
organism, 101
oscillating, 115
Overexploiting, 13
overpopulation, 102
Pakistan, 17
paradigm, 60
particles, 9, 10, 60, 61
Paul H. Ray, 59
people, 11, 16-8, 31, 34, 42, 55-7, 59, 112, 114, 128
pests, 6, 7, 9
philanthropists, 156, 157
philanthropy, 157
philosophers, 33-5, 117, 119
photosynthesis, 102, 141
pine beetles, 7
planetesimals, 100
planets, 36, 37, 47, 75, 99, 100, 108
plankton, 10, 14
planned obsolescence, 23
plastic, 10

Pol Pot, 16
polluters, 10
pollution, 9, 10
poor, 18
population, 11, 13-6, 22, 23, 29, 30, 31, 34, 42, 59, 113, 119, 127, 128, 141, 144, 145, 158, 159, 160, 163, 164, 167, 168
positive, 19
postmodern stage, 58
potential, 7, 75, 97, 151
poverty, 16, 23, 148, 160, 162, 167
predictions, 136
pre-history, 33
probability fields, 61
problems, 4, 9, 14, 19, 21, 24, 26-32, 57, 74, 75, 119-21, 152
projection, 14, 144
Prokaryotes, 101
quadrants, 83, 85, 88, 90
quantum field, 97
quantum physics, 60, 61
random mutations, 102
reductionist, 49
refugees, 6
religions, 28-30, 39, 40, 42, 114
religious, 29, 33-5, 38-40, 42, 59, 114
renaissance, 4
reorganization, 54, 72, 85, 121, 149, 177
reproduction, 31, 101, 104, 167
resources, 13-5, 23, 114, 143

revolutionary, 47, 151, 152, 155, 157, 160, 177, 178
rich, 18, 114
risk, 9, 27
Roman Empire, 33, 34
Salman Khan, 162
San Francisco, 8
sanitation, 164
sapiens, 106, 112, 152
scientific method, 37, 114
scientists, 35, 48, 112
S-curve, 140, 141, 143, 144, 146, 150
sea level, 7
self-adjust, 64
self-adjusting, 65
self-organize, 51
self-organizing, 51, 54, 57, 84, 85, 95
Sherry Ruth Anderson, 59
shortage, 13, 15, 102, 140
singularity, 97, 142
small change, 65, 66, 68
social, 13, 17, 48, 56, 84, 89-91, 113, 114
soil, 9
solar cells, 165, 166
space, 47, 60, 97, 99
Spiral Dynamics, 57, 115
stability, 55, 64, 68, 94, 113
stages, 83, 89, 112
star birth, 99
stars, 34, 36, 61, 75, 99, 100, 108, 142
starvation, 13, 16, 17, 24
states, 83, 90
status quo, 66, 67, 136

Steve McIntosh, 57
storage capacity, 131
strange attractor, 52
strange attractors, 52, 77, 149
stress, 17, 54, 64-75, 77, 94, 95, 102, 117, 118, 120-2
stresses, 54, 64, 90
subconscious, 174
Sugata Mitra, 161
sun, 34, 36, 99, 100, 102
Sun, 61
supernova, 99, 142
superstition, 28, 35, 112
superstitious, 34
survival of the fittest, 94, 96, 113
sustain, 11, 13
sustainability, 15, 22, 150
Sustainability, 148
sustainable, 13, 14-6, 29, 30, 102, 152
symbiosis, 96
symbiotic relationships, 103, 171
symbolic thinking, 111
system theory, 49, 50, 56
technology, 23, 30, 151
terrorism, 16, 18
thermonuclear fusion, 99, 142
tier 1, 118, 147, 152
tier 2, 118, 147, 152
time, 60, 97, 98, 141
toilet, 164
tornadoes, 6
traditional, 29, 42, 59, 117, 149

traditional stage, 58
transformational change, 67, 68, 169
transition, 34, 68, 69, 72, 74, 77, 113, 117, 118, 150
trash, 22
Traveling Wave Reactor, 166
trends, 40, 108, 109, 111, 115, 121, 123, 127, 139, 143, 144, 146-9, 155, 170, 172, 174
tribal, 58, 59, 112, 113, 117
truancy, 162
tutor, 163
TWR, 166, 167
types, 83, 90
tyranny, 17
underdeveloped, 13, 149, 150
uneducated, 161, 167
unhealthy, 26, 58
United States, 22, 40
universe, 34, 36, 37, 60, 61, 68, 69, 93, 96-9, 107, 108
unrest, 13, 15
unsustainable, 15, 22
Urbanization, 34
violations, 10
violence, 16, 17, 28, 29
volcanoes, 100
warmer, 7
Warren Buffett, 157
warrior, 59, 113, 117
warrior stage, 58

waste, 9, 14, 31, 101, 102, 103
water, 7, 9, 10, 13-5, 18, 31, 83, 100, 158, 164
weather, 6, 7, 51-3, 140
wildfires, 6
wildlife, 10
world view, 34, 41, 42, 49, 59, 91, 119, 148
Yosemite, 7
zealots, 17, 18

About the Author

Robert B. Calkins is a retired aerospace research and development systems engineer. His forty year career in aerospace was spent working for the U.S. Air Force, McDonnell Douglas and the Boeing Company. His education included a B.S. degree in Aerospace Engineering, a B.A. in Applied Mathematics, an M.S. in Computer Science and a MA. During his carrier, he published a number of technical papers, reports, journal articles and authored a national technical standard. He was also active in several technical societies and holds a couple of patents.

Robert lives with his wife in the woods in the Pacific North West. He has been doing research for several years on the systems approach to understanding our world, society and current conditions. His hobbies are hiking, photography and graphic art. Past hobbies included skydiving and general aviation.

https://www.amazon.com/author/robertbcalkins

Made in the USA
San Bernardino, CA
13 November 2016